Complex manifold techniques in theoretical physics

Research Notes in Mathematics
Sub-series in Mathematical Physics

D E Lerner & P D Sommers (Editors)

University of Kansas & North Carolina State University at Raleigh

Complex manifold techniques in theoretical physics

Pitman Advanced Publishing Program

SAN FRANCISCO · LONDON · MELBOURNE

FEARON PITMAN PUBLISHERS INC.
6 Davis Drive, Belmont, California 94002

PITMAN PUBLISHING LIMITED
39 Parker Street, London WC2B 5PB
North American Editorial Office
1020 Plain Street, Marshfield, Massachusetts 02050

Associated Companies
Copp Clark Pitman, Toronto
Pitman Publishing New Zealand Ltd, Wellington
Pitman Publishing Pty Ltd, Melbourne

AMS Subject Classifications: 81-XX, 14-XX, 53-XX

Manufactured in Great Britain

ISBN 0 273 08437 2

Preface

Over the past few years, a remarkable relationship between complex analysis and mathematical physics has emerged. Through the use of Penrose's twistor theory, the solutions to some fundamental differential equations of mathematical physics can be converted into complex analytic objects on projective space. The corresponding classification problems of complex geometry can, in principle, be solved completely, with the result that one now has an explicit description of the set of all solutions to the (self-dual) Yang-Mills, Einstein, and zero rest-mass equations.

The articles in this volume are based on talks presented at a workshop on this subject held at Lawrence, Kansas from July 10 through July 15, 1978. The principal speakers at the workshop were D. Burns (Princeton), R. Hartshorne (Berkeley), R. Jackiw (M.I.T.), E. T. Newman (Pittsburgh), R. Penrose (Oxford), R. S. Ward (Oxford), and R. O. Wells, Jr. (Rice). In addition to the articles by the principal speakers, these proceedings also contain several articles of a more technical nature contributed by the workshop participants.

The workshop was supported by the National Science Foundation under the auspices of the Rocky Mountain Mathematics Consortium. The workshop organizers were D. Lerner and P. Sommers.

Contents

Introduction 1

R. Jackiw

Non-linear equations in particle physics 5

R. S. Ward

The self-dual Yang-Mills and Einstein equations 12

R. Hartshorne

Algebraic vector bundles on projective spaces, with applications to the Yang-Mills equation 35

N. H. Christ

Self-dual Yang-Mills solutions 45

R. Penrose

On the twistor descriptions of massless fields 55

R. O. Wells, Jr

Cohomology and the Penrose transform 92

L. P. Hughston

Some new contour integral formulae 115

C. M. Patton

Zero rest mass fields and the Bargmann complex structure 126

D. Burns

Some background and examples in deformation theory 135

E. T. Newman

Deformed twistor space and H-space 154

K. P. Tod

Remarks on asymptotically flat H-spaces 166

J. Isenberg and P. B. Yasskin

Twistor description of non-self-dual Yang-Mills fields 180

E. Witten

Some comments on the recent twistor space constructions 207

J. Harnad, L. Vinet and S. Shnider

Solutions to Yang-Mills equations on $\overline{M}^4$ invariant under subgroups of O(4,2) 219

P. Green

Integrality of the Coulomb charge in the line space formalism 231

N. H. Christ

Unified weak and electromagnetic interactions, baryon number nonconservation and the Atiyah-Singer theorem 234

Introduction

The differential equations of classical and quantum physics have provided a fertile common background for collaboration between mathematicians and physicists for many years. In particular, the desire to understand linear differential equations has led the mathematician to significant advances in the areas of functional analysis, differential and complex geometry, and the theory of Lie groups. Many of these same developments have given the physicist a deeper understanding both of the successes of the quantum theory and of the problems remaining to be solved.

In the case of non-linear equations, however, the results have been less spectacular. This is particularly unfortunate from the physicists' standpoint, for these are the equations related to the description of non-trivial interactions. A case in point at the classical level is that of general relativity, in which the gravitational field is coupled to itself through the non-linear terms in Einstein's equations: On the one hand, there are the short-time existence and uniqueness theorems for the Cauchy problem; on the other, a finite number of families of exact solutions, many of which appear to be devoid of physical significance. There is no known method of constructing the general local solution from free data. The only information available as to the qualitative nature of global solutions suggests that, in the generic case, singularities develop and the solutions do not persist for all time. It goes without saying that our poor understanding of the classical fields does not make the quantum mechanical problem any easier.

Einstein's equations illustrate several typical problems which arise whenever one is confronted with a system of non-linear partial differential equations:

(a) What is the nature of a generic local solution? How does one construct solutions explicitly?

(b) What can be said concerning the set of all solutions (possibly subject to some suitable boundary conditions)? Can this set even be defined in a rigorous way, and, if so, how is it parametrized?

These are extremely difficult questions, and it is a remarkable discovery that the mathematical techniques routinely employed by algebraic geometers and complex analysts can, in some cases, be brought to bear on these problems with stunning effect. In several important instances, the solutions to a given system of field equations can be represented entirely in terms of complex analytic objects. Very roughly speaking, the field equations can be converted into the Cauchy-Riemann equations by making suitable changes in the geometric background space.

There are, at the moment, three concrete examples of this phenomenon:

Massless free fields: The positive frequency fields of a definite helicity have been shown by Penrose to be isomorphic to certain sheaf cohomology groups over a fixed open subset of projective three-space, $P_3(C)$. Any real-analytic field can be similarly represented, and non-analytic fields can be represented in terms of relative cohomology groups and hyperfunctions, as shown by Wells.

Yang-Mills fields: Ward has shown that any real-analytic self-dual Yang-Mills field defined in an open region of Minkowski space determines a holomorphic vector bundle on an open subset of $P_3(C)$. The original field can be reconstructed from the transition functions of the bundle. The (Euclidean)

SU(2) Yang-Mills fields describing pseudoparticles or instantons are classified by stable rank-2 vector bundles on $P_3(C)$ satisfying a certain reality condition. The n-instanton solutions are parametrized by the points of a real algebraic variety of dimension 8n-3 (Atiyah, Hitchin, Singer, and Ward).

Half-flat solutions to Einstein's equations: These are (necessarily complex) four-manifolds with Ricci-flat holomorphic metrics whose curvature tensors are either self-dual or anti-self-dual. Penrose has shown that any local solution sufficiently close to flat space can be obtained by deforming the complex structure of a neighborhood of a line in $P_3(C)$ in an appropriate way. Starting from an altogether different point of view, Newman has shown how to associate a half-flat spacetime with any real-analytic asymptotically flat vacuum solution sufficiently close to Minkowski space. In this context, the half-flat spacetime is called an H-space. Newman's construction also relies on deformation theory, and the two approaches are closely related.

While these examples differ significantly in their details, in each case the set of solutions is in one-to-one correspondence with a well defined class of complex analytic objects. The differential equations themselves have disappeared, having been in some sense coded into pure complex structure; and the problem of enumerating and classifying the solutions has been converted into a problem in complex geometry (e.g., classifying a subset of the stable rank-2 vector bundles).

Since these results are quite recent, only the barest outlines of what may eventually become a systematic and powerful theory are understood at the moment. Even so, the subject appears quite promising for the following reasons:

(1) The physical problems which have been successfully handled so far are non-trivial and of considerable current interest. Moreover, the techniques involved in the constructions were developed, not for their mathematical generality, but to deal specifically with problems of physical significance. It is reasonable to expect that further developments in the theory will have direct physical applications. It is even possible that some of the long standing problems of quantum field theory may be solved by a reformulation involving complex manifolds.

(2) From the mathematician's point of view, this is a largely unexplored area with great potential. The examples cited above indicate the existence of a class of differential equations which can be effectively treated by complex manifold techniques. It is perhaps worth emphasizing three of the common features of these examples:

(a) The methods are constructive--given the appropriate complex analytic object, one can in principle, and frequently in practice, proceed to construct a solution in closed form.

(b) The general local solution is obtained in each case.

(c) Certain types of boundary conditions, such as finite action or positive frequency, can be imposed on the global solutions by properly restricting the class of complex analytic objects.

It is clearly an important matter to determine the full generality of these methods.

D E LERNER
Department of Mathematics
University of Kansas
Lawrence, Kansas 66045

P SOMMERS
Department of Mathematics
North Carolina State University
Raleigh,
North Carolina 27650

R Jackiw

Non-linear equations in particle physics

In the last few years, a new approximation procedure has been developed for analyzing quantum field theory. This method has allowed physicists who are examining various field theoretical models to get results which had not been anticipated from earlier, perturbative investigations. The new approach begins with a study of the relevant field equations, viewed as partial differential equations for ordinary functions (i.e. for classical fields) rather than as Heisenberg field equations for quantum operator fields. The field equations are analyzed and solved by techniques of mathematical physics, and then the quantal information is extracted from these classical solutions. In other words, quantum field theorists have learned what quantum mechanicians of the 1920's knew already: semi-classical methods can give an approximate quantal description of a physical system, provided the classical description of that system is completely known; this is just as true for modern quantum field theory as it is for particle quantum mechanics.

In order to make progress with the new program, physicists have been solving partial, non-linear differential equations which arise in this context. It happens that these equations are also mathematically significant; hence we are witnessing a fortunate conjunction of interest between physicists and mathematicians, and that is why a physicist is here addressing an audience of mathematicians.

Several reviews of the subject are available; consequently this written record of my lecture will be brief and the interested reader is referred to the following extensive summaries.

S. Coleman, "Classical Lumps and their Quantum Descendants" in New Phenomena in Sub-Nuclear Physics, edited by A. Zichichi, Plenum Press, New York (1977).

S. Coleman, "Uses of Instantons" in N. N. edited by A. Zichichi, Plenum Press, New York (in press).

R. Jackiw, "Quantum Meaning of Classical Field Theory" Reviews of Modern Physics 49, 681 (1977).

R. Jackiw, C. Nohl and C. Rebbi, "Classical and semi-Classical Solutions of the Yang-Mills Theory" in Particles and Fields, edited by D. Boal and A. Kamal, Plenum Press, New York (1978).

A very short account is by

R. Jackiw and C. Rebbi, "Topological Solitons and Instantons" Comments on Nuclear and Particle Physics (in press).

Of course the partial differential equations possess a vast variety of solutions, and only certain special ones have been used thus far by physicists to probe the quantum theory. These are the "soliton" solutions and the "pseudoparticle" or "instanton" solutions. Here I shall describe them and explain their significance for physical theory.

The solitons are real solutions with finite energy and localization in space. Also they are stable against small deformations. (This definition does not coincide with usage in the engineering and applied mathematics literature.) Their behavior at large spatial distances frequently is topologically non-trivial; in that case, no finite-energy evolution can connect a solution with the soliton's asymptotes to a solution with the vacuum's asymptotes. There is an infinite-energy barrier separating the two. (The "vacuum" is the zero-energy solution of the equations.) For the quantum theory, these classical soliton solutions are interpreted as evidence that there exist unexpected particle states, the so-called "quantum solitons," which are neither the "elementary" particles of the model, nor are they bound states of a finite number of elementary particles. Rather quantum solitons are bound states of an infinite number of elementary particles; they represent collective, coherent excitations. Quantum soliton states arise in various field theoretical models, but at present it is not known what phenomenological significance should be assigned to them. The theory of soliton solutions is discussed in the first and third reviews cited above.

Instantons (pseudoparticles) are real solutions of field equations that are continued to imaginary time ($t \to -i\tau$). It has been established that such solutions can be used to give an approximate, semi-classical description of quantum mechanical tunnelling. Numerical computation of the tunnelling amplitude proceeds according to the following sequence of steps. Firstly the quantum field theory is formualted in terms of n-point functions (vacuum expectation values of the time-ordered products of n quantum fields). These n-point functions can be obtained from a generating functional for which there is a functional integral representation. Next the functional integral is continued from Minkowski space to Euclidean space. Finally, the Euclidean space functional integral is evaluated approximately by expanding about the Euclidean (imaginary-time) solutions.

Instanton solutions and calculations of tunnelling have been principally performed for Yang-Mills theory--the field theory which is currently believed to be appropriate for a correct description of fundamental processes in Nature. A phenomenologically important result has thus been achieved: tunnelling reduces the degree of symmetry so that physical predictions of the theory possess less symmetry than had been anticipated. Precisely such a reduction of symmetry is required to obtain agreement with experimental data. The second and fourth reviews mentioned above are devoted to this topic.

Yang-Mills instanton solutions can be labelled by their Pontryagin index. If A_a^μ is an SU(2) gauge potential in Euclidean space (a = 1, 2, 3, internal symmetry indices; μ = 1, 2, 3, 4, Euclidean space indices) and $F_a^{\mu\nu}$ is the corresponding gauge field, then the Pontryagin index is $q = \frac{1}{32\pi^2}\int d^4x\, {}^*F_a^{\mu\nu} F_{a\mu\nu}$, where ${}^*F_a^{\mu\nu}$ is the dual gauge field, related to the gauge field by the anti-symmetric tensor, ${}^*F_a^{\mu\nu} = \frac{1}{2}\,\varepsilon^{\mu\nu\alpha\beta} F_{a\alpha\beta}$. For sufficiently well-behaved fields, q is an integer, and for the known instanton solutions it takes on arbitrary integral values. Moreover, these solutions are self-dual (${}^*F_a^{\mu\nu} = \pm F_a^{\mu\nu}$). Both physicists and mathematicians have classified all self-dual instantons and have shown that when q = n, the solution depends on $8|n|-3$ gauge invariant parameters (for the SU(2) group; analogous formulas hold for arbitrary groups). The physical interpretation of these parameters is as follows: the q = n instanton can be thought to be a non-linear super-position of $|n|$ $q=\pm 1$ instantons. Each instanton requires 8 parameters to specify it: 4 determine the location in Euclidean space, 1 parameter sets the instanton's size, while the remaining 3 give its orientation in group space. The total $8|n|$ is decreased by 3, which represents an overall gauge rotation of the entire ensemble. Physicists have given explicitly solutions with $5|n|+4$ parameters; a constructive procedure for obtaining the most general $8|n|-3$ parameter solution has been found by mathematicians, and is described in this conference by Ward.

While these results have shed much light on the mathematical structure of Yang-Mills theory, their physical import has not as yet been fully estab-lished. Only the $|n| = 1$ solution has played a role in practical, physical calculations; those with $|n| > 1$ appear to give only negligibly small corrections. Thus one open question concerns the proper role of instantons with $|n| > 1$ in physical theory. Other interesting questions which remain as yet unanswered include the following. What are the physically relevant regularity conditions at infinity which render the action ($\equiv \frac{1}{4}\int d^4x\, F_a^{\mu\nu} F_{a\mu\nu}$) finite and the Pontryagin index q integer-valued? (The known instantons have finite action and integer-valued q.) What distinguishes solutions of Yang-Mills equations that are not self-dual and do they have significance for physical theory? (Mathematicians have recently shown that stable solutions are necessarily self-dual.) What are the properties of the $8|n|-3$ parameter space of solutions and does there exist a group of symmetry transformations acting on these parameters? (For $|n| = 1$ the answer is known. The 15 parameter conformal group O(5,1)--a symmetry group of the Yang-Mills theory--acts on the 5 parameter $|n| = 1$ solution as follows: the 10 parameter O(5) subgroup leaves the solution invariant; the remaining 5 parameter family of transformations modifies the parameters of the solution.) Some progress on this question is here reported by Hartshorne.

Finally, impressed by the marvelous structure of Euclidean Yang-Mills theory, some of us have returned to the Minkowski space theory and looked for classical solutions there. It is known that no soliton (non-dissipative, particle-like) solutions exist, but other kinds of solutions have been found. It has proved useful to classify them according to their transformation properties under the Minkowski-space conformal group, O(4,2), and its subgroups. Especially interesting are real solutions invariant under the O(4) × O(2) subgroup, as well as those invariant under the O(4) subgroup. Complex solutions, invariant under other subgroups, have also been exhibited and are described at this meeting by Shnider. It has been suggested that the real solutions may be used to determine the semi-classical spectrum of various conformal generators (see the fourth reference). The role of complex solutions in the physical theory is not known.

Another body of calculation that has interested physicists and mathematicians concerns the coupling of an external, linear system to a soliton or instanton; for example one studies the Dirac or Klein-Gordon equations in the field of an instanton or soliton. In this way one is led to examine linear, elliptic differential operators which frequently possess vanishing eigenvalues. (Small deformations of solitons or instantons satisfy equations of this type as well.) The zero-eigenvalue modes have profound significance for the physical theory; their role is explained in the third and fourth cited reviews. Close study of these modes has also put physicists in contact with mathematicians since the principal mathematical result relevant here is the Atiyah-Singer Index Theorem or equivalently, as physicists came to know it, the anomaly of the axial vector current.

The semi-classical method, based on classical solutions, has provided new understanding of quantum field theory. However, we must remember that a complete reconstruction of the quantum theory from classical solutions is not in general possible. Certainly the restriction to finite energy or finite action, which has characterized the solutions emphasized by all of us thus far, is only appropriate to approximate quantum mechanical calculations; classical solutions with infinite energy and/or action are also needed for understanding the full quantal system. Moreover, recall that a functional integral representation of a quantum theory involves a functional integration over all classical configurations, not just those satisfying equations of motion. Hence, in general, we shall have to go beyond classical solutions and examine classical field configurations to obtain further insights into the quantum theory.

Acknowledgments

This summary of my lecture was prepared at the Aspen Center for Physics and is based on notes taken by Professor B. DeFacio, whose help is gratefully acknowledged. The research is supported in part through funds provided by the U.S. Department of Energy (DOE) under contract EY-76-C-02-3069.

R JACKIW
Center for Theoretical Physics
Laboratory for Nuclear Science and
Department of Physics
Massachusetts Institute of Technology
Cambridge, Massachusetts 02139

R S Ward

The self-dual Yang-Mills and Einstein equations

§1. INTRODUCTION

As was pointed out by Jackiw in his lecture, it is often useful to study the classical (i.e. non-quantum) versions of quantum field theories. One such classical system is Yang-Mills theory. Another classical theory (which does not, as yet, have a satisfactory quantum version) is Einstein's theory of gravity. The Yang-Mills equations and Einstein's equations are hyperbolic systems of non-linear equations in space-time. In this lecture we shall be working with their analytic extension to *complexified* space-time and, in particular, to positive definite 4-space (so that they become elliptic rather than hyperbolic).

The Yang-Mills equations and Einstein equations, being non-linear, are by no means easy to solve. This is where complex manifold techniques come in. When one transforms the problem from space-time to a certain complex manifold called *twistor space*, one discovers that the *self-dual* solutions of the Yang-Mills and Einstein equations can all be obtained without having to solve any differential equations at all. This transformation (the so-called Penrose transform) is the subject of the present lecture.

Twistor theory was invented and developed by R. Penrose; in particular, it was he who produced the crucial theorems 2.3 and 5.1 discussed below. The application described in §§3 and 4 of twistor formalism to Yang-Mills theory is due mainly to M.F. Atiyah and myself.

To begin with, we need to establish some notation. Let R denote the real numbers, C the complex numbers and C* the multiplicative group of nonzero complex numbers. *Complexified Minkowski space-time CM* is C^4 equipped with a flat, nondegenerate, holomorphic metric g. [In other words, there exist complex coordinates x^a (a = 0,1,2,3) on CM such that g is given by the *complex line element*

$$ds^2 = \sum_a (dx^a)^2.]$$

Clearly CM can also be thought of as the complexification of Euclidean 4-space E^4. Let T denote the holomorphic tangent bundle of CM and Λ^n the bundle of holomorphic n-forms. The tangent space at $x \in CM$ is denoted T_x, and so forth.

Choose a constant volume element (4-form) on CM; this, together with the metric, defines a *duality operator*

$$* : \Lambda^2 \to \Lambda^2,$$

with $** =$ identity. Then Λ^2 may be decomposed into orthogonal subspaces:

$$\Lambda^2 = \Lambda^{2+} \oplus \Lambda^{2-},$$

where $\Lambda^{2\pm}$ is the eigenspace of $*$ corresponding to the eigenvalue ± 1. The sections of Λ^{2+} are called *self-dual* 2-forms, while those of Λ^{2-} are *anti-self-dual*.

Now let us introduce spinors. There are two basic spin-bundles, denoted S and S'; their duals are denoted S* and S'* respectively. Each of them is a 2-complex-dimensional vector bundle over CM with structure group SL(2,C); this means that each fibre S_x, $x \in CM$, is equipped with a symplectic form ε.

Spinors are tied to the background geometry in the following way. First, one chooses an isomorphism

$$T \xrightarrow{\cong} S \otimes S'. \qquad (1.1)$$

The dual version of this is

$$\Lambda^1 \xrightarrow{\cong} S^* \otimes S'^*. \tag{1.2}$$

We require that the metric g and the symplectic forms ε and ε' on S and S' be related according to

$$g = \varepsilon \otimes \varepsilon', \tag{1.3}$$

where the equality sign is to be understood in terms of the isomorphism (1.2). It follows from this a vector v in T_x is *null* (i.e. has zero length) iff it is decomposable in $S_x \otimes S'_x$. In other words,

$$g(v,v) = 0 \Longleftrightarrow v = \lambda \otimes \pi, \tag{1.4}$$

where $\lambda \in S_x$, $\pi \in S'_x$.

Let S(n) denote the symmetrized tensor product of n copies of S, and similarly for S'(n), S*(n) and S'*(n). From (1.2) and the two-dimensionality of spin-space it follows that

$$\Lambda^2 \cong S^*(2) \oplus S'^*(2).$$

This is the decomposition of Λ^2 into self-dual and anti-self-dual parts — but we still have to choose which part is which. We make the choice

$$\Lambda^{2+} \cong S'^*(2), \tag{1.5}$$

$$\Lambda^{2-} \cong S^*(2). \tag{1.6}$$

Each spin-bundle is equipped with a connection

$$\nabla : S \to S \otimes \Lambda^1, \tag{1.7}$$

etc. In CM all these connections are *flat*.

Complex projective m-space is denoted P_m. Let $L(1)$ be the hyperplane section bundle over P_m and let $L(n)$ be the line bundle $[L(1)]^n$. Let $\mathcal{O}(n)$ denote the sheaf of germs of holomorphic sections of $L(n)$. (So when written in terms of homogeneous coordinates, sections of $\mathcal{O}(n)$ correspond to holomorphic functions homogeneous of degree n.) Since $L(0)$ is the trivial bundle, $\mathcal{O}(0)$ is the same as $\mathcal{O}$, the sheaf of germs of holomorphic functions on P_m. One final bit of notation: if E is a holomorphic vector bundle, then $\Gamma(E)$ denotes the group of holomorphic sections of E.

§2. TWISTOR GEOMETRY AND MASSLESS FREE FIELDS

We begin with a description of the geometrical correspondence between complexified space-time and twistor space; for more details, see [9].

An α-*plane* is defined to be a complex 2-plane Z in CM with the following two properties. Let u and v be any two vector fields on Z, tangent to Z; let μ and ν be the corresponding 1-forms. Then

(i) Z is *totally null* (i.e. isotropic): $g(u,v) = 0$;

(ii) Z is *self-dual*: $\mu \wedge \nu$ is a self-dual 2-form.

Similarly, a β-*plane* is a totally null *anti*-self dual 2-plane in CM. In what follows we shall be using β-planes rather than α-planes. But every statement has an "α-version" in which "self-dual" and "anti-self-dual" are interchanged.

Proposition 2.1 The space of β-planes in CM is the quotient of $S'_o \times (S^*_o - \{0\})$ by the natural multiplicative action of C^*.

Proof S'_o etc. means spinors at the origin: we are using the flat connection on CM to refer everything to the origin. A point $x \in$ CM is represented by its position vector with respect to the origin:

$$x \in T_o = S_o \otimes S'_o = \mathrm{Hom}\,(S^*_o, S'_o).$$

So given $(\omega,\pi) \in S'_o \times (S^*_o - \{0\})$, the equation

$$\omega = x.\pi \qquad (2.1)$$

makes sense. For fixed (ω,π), the solution space of (2.1) in x is a β-plane (this follows from equations (1.4) and (1.6)). Every β-plane arises in this way; (ω,π) and $(\tilde{\omega},\tilde{\pi})$ determine the same β-plane iff

$$\omega = \lambda\tilde{\omega}, \quad \pi = \lambda\tilde{\pi} \quad \text{for some } \lambda \in C^*. \quad \square$$

Now $(S'_o \times S^*_o)/C^* \cong C^4/C^* = P_3$, so Proposition 2.1 tells us that the space of β-planes in CM is $P_3 - I$, where I is the line in P_3 given by $\pi = 0$. Here (and in what follows) "line" means "complex projective line"; so intrinsically $I \cong P_1$.

We say that $P_3 - I$ is the *projective twistor space* corresponding to CM; each β-plane Z corresponds to a *projective twistor* $\hat{Z} \in P_3 - I$. [In the literature these are usually called *dual* projective twistors, while the α-planes correspond to projective twistors.] Next we consider what *space-time points* correspond to in twistor space. A point $x \in CM$ is characterized by the set of all β-planes through it: from equation (2.1) it follows that this set is the *line*

$$\hat{x} = \{(\omega,\pi)\,|\,\pi \neq 0, \;\; \omega = x.\pi\}$$

in $P_3 - I$. So we can summarize the space-time ↔ twistor space correspondence as point $x \in CM$ $\leftrightarrow$ line $\hat{x}$ in $P_3 - I$,

β-plane Z in CM $\leftrightarrow$ point $\hat{Z} \in P_3 - I$.

The space of all lines in P_3 is a compact 4-dimensional complex manifold, in fact a quadric Q in P_5. So CM (thought of as the space of lines in $P_3 - I$) is an open submanifold of Q; and Q is a *compactification* of CM.

The crucial feature of the above correspondence is the way that it relates to the *conformal* (i.e. null-cone) structure of CM: the following result is easily verified.

<u>Proposition 2.2</u> Two points $x, y \in CM$ lie on a common null geodesic iff the corresponding lines $\hat{x}$ and $\hat{y}$ in $P_3 - I$ intersect.

In this lecture we shall be working with two variants of the CM $\leftrightarrow$ $P_3 - I$ correspondence, namely the *local* version and the *global positive definite* version. In the local version we use an "open ball" X in CM. By this is meant that X is an open submanifold of CM, such that the intersection of X with any linear subspace of CM is topologically and holomorphically trivial. The corresponding region $\hat{X}$ of twistor space is defined to be the space of β-planes which intersect X. So, in other words, we have the correspondence

point $x \in X$ $\leftrightarrow$ line $\hat{x}$ in $\hat{X}$,

β-plane Z in X $\leftrightarrow$ point $\hat{Z} \in \hat{X}$.

The space $\hat{X}$ has the topology $S^2 \times R^4$ (as opposed to X, which is of course topologically trivial).

For the positive definite version we restrict our attention to a Euclidean "slice" E^4 of CM. The structure group on the spin-bundles over E^4 is reduced from SL(2,C) to SU(2). This means that each spin-space S is equipped with a complex structure $\sigma: S \to S$, such that

$\varepsilon(.,\sigma.)$

is a positive definite Hermitian form.

Now σ induces an action on $T = S \otimes S'$, the fixed points of which are precisely the real vectors, i.e. the real points E^4 in CM. It also gives rise to an anti-holomorphic involution, again denoted σ, on $P_3 = (S' \times S^*)/C^*$ and on $P_3 - I$. This map $\sigma: P_3 \to P_3$ has no fixed points. However, it does have fixed lines: let us call these "real" lines. Thus

the real points $x \in E^4$ *correspond to the real lines* $\hat{x}$ *in* $P_3 - I$.

In addition, the line I is a real line, so the real lines in P_3 correspond to the points of the one-point compactification S^4 of E^4. Clearly no two distinct real lines can intersect each other (if they did, then the corresponding points on S^4 would be null-separated; but this is impossible, since the metric on S^4 is positive definite). So P_3 *is a bundle over* S^4, with the real lines as fibres.

It should be remarked that real Minkowski space-time (signature +---) can be obtained by imposing a different "real structure" on CM and P_3. The spin-bundles S and S' remain SL(2,C)-bundles, but become complex conjugates of each other. The upshot of this is that every self-dual real 2-form in Minkowski space-time is also anti-self-dual, and hence vanishes.

Finally in this section we discuss massless free fields. Equations (1.2) and (1.7) imply that the derivative operator ∇ maps sections of $S^*(n)$ to sections of $S^*(n) \otimes S^* \otimes S'^*$. Now $S^*(n) \otimes S^*$ decomposes into its irreducible parts as follows:

$$S^*(n) \otimes S^* = S^*(n+1) \oplus S^*(n-1).$$

Let τ be the projection from $S^*(n) \otimes S^* \otimes S'^*$ onto $S^*(n+1) \otimes S'^*$ and ζ the projection from $S^*(n) \otimes S^* \otimes S'^*$ onto $S^*(n-1) \otimes S'^*$. Then

$$\tau o \nabla : \Gamma(S^*(n)) \to \Gamma(S^*(n+1) \otimes S'^*) \tag{2.2}$$

is called the *twistor operator*, and

$$\zeta o \nabla : \Gamma(S^*(n)) \to \Gamma(S^*(n-1) \otimes S'^*) \tag{2.3}$$

is called the *Dirac operator*. Both these operators have "primed" versions; for example, there is a primed Dirac operator

$$\zeta' o \nabla : \Gamma(S'^*(n)) \to \Gamma(S'^*(n)) \to \Gamma(S'^*(n-1) \otimes S^*).$$

A massless free field is a symmetric spinor field which is annihilated by the Dirac operator. More precisely, let X be an open ball in CM and let n be any integer. Then the space $\Phi_n(X)$ of *helicity* $\frac{1}{2}n$ *massless free fields* in X is

(i) if $n > 0$, the kernel of $\zeta' o \nabla$ acting on sections of $S'^*(n)$ restricted to X;

(ii) if $n < 0$, the kernel of $\zeta o \nabla$ acting on sections of $S^*(-n)$ restricted to X;

(iii) if $n = 0$, the kernel of $\Box$ (the covariant Laplacian/d'Alembertian) acting on scalar fields in X.

<u>Theorem 2.3</u> The vector spaces $\Phi_n(X)$ and $H^1(\hat{X}, \mathcal{O}(n-2))$ are naturally isomorphic.

<u>Remark</u> For a proof, see Penrose's article in these proceedings. The theorem developed out of the well-known *Kirchoff integral formula* for massless free fields. This integral formula (in *real* Minkowski space-time) expresses the value of the massless field at the point x as an integral of the initial data over some cross-section of the past null (i.e. characteristic

cone of x. In general, for a hyperbolic equation, no such formula exists, because there is usually a contribution from *inside* the past characteristic cone. The existence of the Kirchoff formula for massless free fields reflects the fact that they satisfy *Huygens' principle* (in fact, this is the usual definition of Huygens' principle).

Among other equations satisfying Huygens' principle are the self-dual Yang-Mills equations and (in some sense) the self-dual Einstein equations. As we shall see, both these equations can be treated very satisfactorily within the twistor formalism. In fact twistor theory seems to have a preference for equations satisfying Huygens' principle: this is hardly surprising, in view of the way in which the space-time ↔ twistor space correspondence is tied up with null-cone (i.e. conformal) geometry of space-time. A corollary of this is that equations which do *not* satisfy Huygens' principle (unfortunately these include the full non-self-dual Yang-Mills and Einstein equations) are likely to be much more difficult to handle. For a discussion of the full Yang-Mills case, see the article by Isenberg and Yasskin in these proceedings.

§3. THE SELF-DUAL YANG-MILLS EQUATIONS

By way of motivation, let us consider a special case of Theorem 2.3, namely the case n = 2:

$$\Phi_2(X) \cong H^1(\hat{X}, \mathcal{O}).$$

Now the isomorphism (1.5) implies that $\Phi_2(X)$ is precisely the space of closed self-dual 2-forms in X — these are called *self-dual Maxwell fields*. The short exact sequence of sheaves

$$0 \to \mathcal{Z} \hookrightarrow \mathcal{O} \xrightarrow{\exp} \mathcal{O}^* \to 0$$

(where Z is the sheaf of integers and $\mathcal{O}^*$ the sheaf of nowhere-zero holomorphic functions) leads to a long exact sequence of cohomology groups, part of which is

$$0 \to H^1(\hat{X},\mathcal{O}) \to H^1(\hat{X},\mathcal{O}^*) \xrightarrow{\gamma} H^2(\hat{X},Z) \to 0.$$

The group $H^1(\hat{X},\mathcal{O}^*)$ is the group of *holomorphic line bundles* over $\hat{X}$ and the map γ maps a bundle to its Chern class in $H^2(\hat{X},Z)$. So we arrive at the following result.

Proposition 3.1 Self-dual Maxwell fields in X correspond in a 1-1 fashion to holomorphic line bundles over $\hat{X}$ with vanishing Chern class.

A Maxwell field, being a closed 2-form in X, is usually interpreted geometrically as the curvature of a connection on a line bundle over X. We can generalize Maxwell theory (and Proposition 3.1) by simply going from line bundles to vector bundles. The resulting theory is called gauge theory or Yang-Mills theory; its basic ingredients will now be summarized.

Let X be an open ball in CM and let E be an r-dimensional complex vector bundle over X. Let A be a linear connection on E: in the language of physics A is a *gauge potential*. If we choose a basis for E (i.e. r linearly independent sections of E), then A may be thought of as an $r \times r$ matrix of 1-forms on X. A section of E is represented as a column r-vector ψ of functions on X, and its covariant derivative is the vector-valued 1-form

$$D\psi := \nabla\psi + A\psi. \tag{3.4}$$

A transformation to a different basis for E is known as a *gauge transformation*. Here the *gauge group* (or structure group) is $GL(r,C)$. One can specialize to some other gauge group $G \subset GL(r,C)$ by requiring that A be a G-connection, i.e. that A be a 1-form with values in the Lie algebra of G.

The *gauge field* F is the curvature of A, represented as a matrix of 2-forms:

$$F = dA + [A,A]. \qquad (3.2)$$

From (3.2) it follows that the curvature F automatically satisfies the Bianchi identities

$$dF + [A,F] = 0. \qquad (3.3)$$

The *Yang-Mills equations* look very similar:

$$d{*F} + [A, {*F}] = 0. \qquad (3.4)$$

Clearly if F is self-dual, i.e. if

$${*F} = F,$$

then the Yang-Mills equations (3.4) follow automatically from (3.3). So the self-dual gauge fields form a subclass of the space of solutions of the Yang-Mills equations.

If Z is a β-plane in X, we can restrict the bundle, connection and curvature to Z:

$$E \mapsto E(Z), \quad A \mapsto A_Z, \quad F \mapsto F_Z.$$

Proposition 3.2 The gauge field F is self-dual iff A_Z is trivial for every β-plane Z.

Proof The connection A_Z on E(Z) is trivial iff its curvature F_Z vanishes. But the β-plane Z is anti-self-dual, so

$$F_Z = 0 \Longleftrightarrow (F^-)_Z = 0,$$

where F^- is the anti-self-dual part of F. Now F^- corresponds, by (1.6), to a spinor $\phi \in S^*(2)$. Using the representation of a twistor Z as a pair of spinors (ω,π), one can show that

$$(F^-)_Z = 0 \text{ for all } Z \Longleftrightarrow \phi(\pi,\pi) = 0 \quad \text{for all } \pi \in S$$
$$\Longleftrightarrow \phi = 0$$
$$\Longleftrightarrow F \text{ is self-dual.} \qquad \square$$

Theorem 3.3 There is a 1-1 correspondence between

(a) connections on the r-dimensional bundle E over X, with self-dual curvature; and

(b) holomorphic r-dimensional vector bundles $\hat{E}$ over $\hat{X}$, satisfying the condition

$$\hat{E}(\hat{x}) \text{ is holomorphically trivial, for all points } x \in X. \qquad (3.5)$$

(Here $\hat{E}(\hat{x})$ means "the bundle $\hat{E}$ restricted to the line $\hat{x}$ in $\hat{X}$".)

Sketch of proof Let us describe geometrically what the correspondence is. To begin with, suppose we are given the bundle E with a self-dual connection. Define K(Z) to be the space of covariantly constant sections of E(Z), i.e.

$$K(Z) = \{\psi \in \Gamma(E(Z)) \,|\, v \lrcorner D\psi = 0 \text{ for all vectors v tangent to Z}\}.$$

Now we define $\hat{E}$ by defining its fibre $\hat{E}_{\hat{Z}}$ over an arbitrary point $\hat{Z} \in \hat{X}$:

$$\hat{E}_{\hat{Z}} \ := \ K(Z).$$

Since $K(Z) \cong C^r$ by Proposition 3.2, it is not difficult to convince oneself that $\hat{E}$ is an r-dimensional holomorphic vector bundle over $\hat{X}$. To establish that $\hat{E}$ satisfies (3.5), it is sufficient to find a biholomorphic isomorphism $\xi:\hat{E}(\hat{x}) \to \hat{x} \times E_x$. Such a map ξ can be defined as follows. Pick $\hat{Z} \in \hat{x}$. A point $\hat{\psi} \in \hat{E}'_{\hat{Z}}$ corresponds by definition to a constant section ψ of $E(Z)$. Now define ξ by

$$\xi : \hat{\psi} \mapsto (\hat{Z}, \psi(x)) \in \hat{x} \times E_x,$$

where $\psi(x)$ means the section ψ evaluated at x.

Conversely, given $\hat{E}$ satisfying (3.5), define E by

$$E_x := \Gamma(\hat{E}(\hat{x})),$$

the space of holomorphic sections of $\hat{E}(\hat{x})$. Since $\hat{E}(\hat{x})$ is trivial and $\hat{x}$ is a Riemann sphere, Liouville's theorem implies that $\Gamma(\hat{E}(\hat{x})) \cong C^r$. So E is an r-dimensional vector bundle over X. It remains to define a self-dual connection on E. We do this by specifying the covariantly constant sections of $E(Z)$, where Z is an arbitrary β-plane. Suppose ψ is a section of $E(Z)$. Thus ψ maps a point $x \in Z$ into $E_x = \Gamma(\hat{E}(\hat{x}))$: in other words, $\psi(x)$ is a section of $\hat{E}(\hat{x})$. Now evaluate this section at $\hat{Z} \in \hat{x}$: this gives $\psi(x)[\hat{Z}] \in \hat{E}_{\hat{Z}}$. Then we say that ψ is a covariantly constant section of $E(Z)$ iff $\psi(x)[\hat{Z}]$ has the same value for all $x \in Z$.

Remarks

(a) Notice that for $r = 1$, Theorem 3.3 reduces to the Maxwell case (Proposition 3.1).

(b) The difference between the space-time and the twistor space pictures is that in the space-time X, the bundle E is trivial and all the information is contained in the *connection* A; in the twistor space $\hat{X}$, however, there is *no* connection: all the information is contained in the (non-trivial) bundle $\hat{E}$.

(c) Theorem 3.3 is a "local" theorem (in the sense of X being an open ball in CM); it has global versions, one of which we shall meet in §4.

(d) It also has versions for gauge groups other than GL(r,C). For example, we could reduce the group to SL(r,C), as follows.

Theorem 3.4 There is a 1-1 correspondence between

(1) self-dual SL(r,C)-connections on E; and

(2) holomorphic bundles $\hat{E}$ satisfying (3.5), as well as the condition

$$\det \hat{E} \text{ is trivial.}$$

(det $\hat{E}$ is the determinant line bundle of $\hat{E}$: its patching function is the determinant of the patching matrix defining the vector bundle $\hat{E}$.)

§4. VECTOR BUNDLES ON P_3 AND INSTANTONS

Instantons (or pseudoparticles) are solutions of the Yang-Mills Equations (3.4) on S^4, with gauge group SU(2) [see Professor Jackiw's lecture]. More general gauge groups are also of interest (e.g. SU(n), Sp(n), etc.), but for simplicity we shall stick to SU(2). The solutions which have been of most interest in connection with physical applications are the self-dual and anti-self-dual solutions (in fact, it is not yet known whether there are any solutions other than these). So what we want is a theorem (analogous to Theorem 3.3) which characterizes self-dual SU(2)-connections over S^4.

Recall from §2 that the anti-holomorphic involution σ acts on P_3 and that the "real lines" of P_3 correspond to the points of S^4.

Theorem 4.1 There is a 1-1 correspondence between

(1) 2-dimensional vector bundles E over S^4 with self-dual SU(2)-connections; and

(2) 2-dimensional holomorphic vector bundles $\hat{E}$ over P_3 satisfying the conditions

(i) $\hat{E}(\hat{x})$ is trivial for all *real* lines $\hat{x}$;

(ii) $c_1(\hat{E})$ (the first Chern class of $\hat{E}$) vanishes;

(iii) σ lifts to an anti-linear map

$$\tilde{\sigma}:\hat{E} \to \hat{E}, \quad \text{with} \quad \tilde{\sigma}^2 = -1.$$

Remarks

(a) For a proof, see [2].

(b) $c_1(\hat{E}) = 0 \Longleftrightarrow \det \hat{E}$ is trivial. In other words (cf. Theorem 3.4), condition (ii) reduces the gauge group from GL(2,C) to SL(2,C). Then condition (iii) achieves the further reduction from SL(2,C) to SU(2).

(c) The only other topological invariant of $\hat{E}$ is $c_2(\hat{E})$: this is called the *instanton number*.

(d) $\hat{E}$ is necessarily algebraic, so all instantons are *rational*.

In his lecture, Hartshorne describes two ways of constructing algebraic vector bundles over P_3. The first is to associate vector bundles with curves in P_3: some of the features of this procedure are discussed below. The second method (due to Horrocks and Barth) will not be pursued further here [see Christ's lecture in these proceedings and reference 1].

The idea of the first method is as follows. Let $\hat{E}$ be a 2-dimensional vector bundle over P_3. Then there exists a positive integer q such that $\hat{E} \otimes L(q)$ has a section which vanishes along a curve Γ in P_3. On $P_3 - \Gamma$, therefore, $\hat{E} \otimes L(q)$ is an extension of the trivial line bundle by another line bundle. After tensoring with L(-q) we arrive at the result that $\hat{E}$, restricted to $P_3 - \Gamma$, is an extension of L(-q) by L(q). Such extensions correspond to elements of the cohomology group $H^1(P_3 - \Gamma, \mathcal{O}(-2q))$ and hence to certain helicity (1-q) massless free fields on space-time. By starting with the massless free fields and working backwards, one can therefore construct all self-dual instantons. This assertion is restated in the following pair of theorems.

Theorem 4.2 Let q be a positive integer and let X be an open ball in CM. Then there exists a non-linear functional A_q from $\Phi_{2-2q}(X) - N$ to the space of self-dual SL(2,C)-connections on the 2-dimensional vector bundle E over X. Here $\Phi_{2-2q}(X)$ is the vector space of helicity (1-q) massless free fields on X and N is a certain subset of this space (see remarks (c) and (d) below).

Theorem 4.3 All self-dual instanton connections A can be obtained in this way. In other words: given A, there exist an integer q, an open ball X in CM and a massless field $\phi \in \Phi_{2-2q}(X)$ such that A (restricted to $X \cap S^4$) is given by $A_q\ [\phi]$.

Remarks

(a) The functionals A_q are described explicitly in [3,4]. This, together with Theorem 4.1 and the algebraic-geometrical results, provides a proof of Theorems 4.2 and 4.3.

(b) Theorem 4.2 is stated in "local" form to emphasize the fact that the A_q are essentially local in nature. (A global version would be somewhat trickier to state, if only because there are no nontrivial massless fields on S^4.)

(c) As an example, let us consider the case q = 1. For more details, see [2]. The metric and volume 4-form on CM define an (essentially unique) linear map

$$f : \Lambda^1 \to \Lambda^1 \otimes \Lambda^{2+}.$$

Identifying Λ^{2+} with the Lie algebra of SL(2,C), we define the functional A_1 by

$$A_1 : \phi \mapsto f(\phi^{-1}\nabla\phi),$$

where $\phi \in \Phi_o(X)$ (i.e. ϕ is a scalar field on X, satisfying $\Box\phi = 0$). Clearly in order for A_1 to be well-defined, we need ϕ to be nowhere zero on X. So the subset N that has to be deleted from $\Phi_o(X)$ consists of those fields which are *somewhere* zero on X. Taking ϕ to be a sum of elementary solutions of $\Box\phi = 0$ based on points of S^4 gives the so-called *'t Hooft instantons*. In this case, the curve Γ is a disjoint collection of real lines in P_3.

(d) For $q > 1$ the picture is similar, but somewhat less straightforward The subset N is again determined by a "somewhere zero" condition. The genus of the curves Γ increases with q [7] so the instanton solutions become more difficult to construct explicitly.

(e) Let M(k) be the space of self-dual instantons with second Chern class $k > 0$. Then M(k) is a real algebraic variety of dimension 8k-3. Let M(k,q) be the subspace of M(k) consisting of those instantons which can be derived from the functional A_q. For each fixed value of k, there exists a value q_k of q such that [7]

$$M(k) = M(k,q_k).$$

But $q_k \to \infty$ as $k \to \infty$; in other words, no finite value of q gives *all* instantons. The following table illustrates these comments.

k	1	2	3	4	5	>5
dim M(k,1)	5	13	19	24	29	5k + 4
dim M(k,2)	5	13	21	29	36	4k + 16
dim M(k)	5	13	21	29	37	8k - 3

Much work remains to be done on investigating the spaces M(k); most recent investigations have concentrated on the Horrocks/Barth approach mentioned above.

§5. SELF-DUAL SOLUTIONS OF EINSTEIN'S EQUATIONS

Let X be a 4-dimensional manifold equipped with a nondegenerate metric g. X may either be real (in which case g is to be real) or complex (in which case g is to be complex holomorphic). In either case, we say that g is *self-dual* if its curvature tensor R is self-dual:

$$*R = R. \tag{5.1}$$

($R \in \Lambda^2 \otimes \Lambda^2$; we dualize R by dualizing either on the first Λ^2 or on the second. By the interchange symmetry of R, it doesn't matter which.) From (5.1) it follows that the Ricci tensor vanishes, so every self-dual metric is a solution of Einstein's vacuum equations.

Recall our comment in §2 that the only real self-dual 2-form in Minkowski space-time is zero. For the same reason, the only real self-dual metric with Lorentz signature is flat (i.e. Minkowski). To escape from this trivial situation, we could either allow g to be complex or require that g be positive definite. Both these cases have applications: for example, *H-space* (which is a complex self-dual space — see Newman's lecture) and *"gravitational instantons"* [8] (which are positive definite).

What we are after is a theorem relating self-dual metrices to "deformed" twistor spaces with a certain structure. To see what this structure is, let us examine the "flat" case of §2 more closely.

Let $\hat{X}$ be the region of P_3 corresponding to an open ball X in CM. Proposition (2.1) implies that

(a) $\hat{X}$ is a holomorphic bundle over P_1. Here P_1 is the projective version of the spin-space S_o^*: in other words, $P_1 = (S_o^* - \{0\})/C^*$. Each fibre of the bundle (a) is a domain in C^2. Let L denote the pullback to $\hat{X}$ of the line bundle $L(1)$ over P_1. Let V be the sub-bundle of the tangent bundle of $\hat{X}$ consisting of the vertical vectors, i.e. of the vectors tangent to the fibres of (a). The second bit of structure we want to single out is

(b) a skew bilinear map $\mu: V \times V \to L^2$. (In the flat case, μ comes from the symplectic form on S'.)

Theorem 5.1 There is a 1-1 correspondence between

(i) sufficiently small self-dual deformations of the metric in X; and

(ii) sufficiently small deformations of $\hat{X}$, preserving the structure (a) and (b).

Remarks

(1) Various bits of the proof may be found spread over a number of references, for example [10,5,6]; the deformation theory of Kodaira is used extensively. An outline of the correspondence will be given below.

(2) The above theorem is a local version: in this context "local" means virtually the same thing as "sufficiently small". Conversely, a global theorem would have to do without the "sufficient smallness". See [2] for a global positive definite theorem. The main difference arising in this case is that some *extra* structure on $\hat{X}$ has to be specified; in the case of Theorem 5.1, this extra structure is guaranteed by Kodaira's deformation theorems and so it doesn't have to be mentioned explicitly.

Let us now describe the correspondence in Theorem 5.1. Suppose first that we are given X with a self-dual metric g. Define a β-*surface* in X to be a complex 2-surface which is totally null and anti-self-dual; then let $\hat{X}$ be the space of β-surfaces in X. The self-duality of g guarantees that $\hat{X}$ is a 3-dimensional complex manifold (as long as g is sufficiently close to being flat) and that the structures (a) and (b) can be defined on $\hat{X}$ in a natural way.

Conversely, let $\hat{X}$ be a deformation satisfying (a) and (b). We want to construct X with a self-dual metric: this is done as follows. X is defined to be the space of *holomorphic sections* of the bundle $\hat{X} \to P_1$. A theorem of Kodaira [6] guarantees (for sufficiently small deformations) that X is a 4-dimensional complex manifold. Suppose that $x \in X$ corresponds to the holomorphic section $\hat{x}$ and let $V(\hat{x})$ be the bundle of vertical vectors restricted to the submanifold $\hat{x}$. So $V(\hat{x})$ is the *normal bundle* of $\hat{x}$ in $\hat{X}$. A vector $v \in T_x$ in X corresponds to a *section* $\hat{v}$ of $V(\hat{x})$. We want to define a metric on T_x. To do so, it is sufficient (cf. Equations (1.1) - (1.3)) to decompose T_x into spin-spaces

$$T_x = S_x \otimes S'_x \tag{5.2}$$

and to define symplectic forms ε_x and ε'_x on S_x and S'_x respectively.

Put $W_x = V(\hat{x}) \otimes L(-1)$ (where $L(-1)$ is thought of as a bundle on the Riemann sphere $\hat{x}$). Thus

$$V(\hat{x}) = W_x \otimes L(1).$$

When we look at sections of $V(\hat{x})$, it turns out that

$$\Gamma(V(\hat{x})) = \Gamma(W_x) \otimes \Gamma(L(1)). \tag{5.3}$$

This is proved [5] by establishing that it holds for the *flat* case and that the various bundles involved are *stable* under small deformations. Now define

$$S'_x := \Gamma(W_x),$$
$$S_x := \Gamma(L(1)). \qquad (5.4)$$

Since $T_x = \Gamma(V(\hat{x}))$, (5.3) and (5.4) give us the desired decomposition (5.2).

The symplectic form ε'_x on S'_x is derived from the skew map

$$\mu: V \times V \to L^2. \qquad (5.5)$$

Restricting (5.5) to $\hat{x}$ and tensoring with L(-1) leads to

$$\mu_x : W_x \times W_x \to L(0)$$

and taking global sections of this gives us

$$\varepsilon'_x : S_x \times S_x \to C,$$

as required. The symplectic form ε on S_x is defined as follows. Going back to the flat case, the symplectic form on S^*_o induces a symplectic form ε on the space $\Gamma(L(1))$ of cross-sections of L(1) over $P_1 = (S^*_o - \{0\})/C^*$. This structure is preserved under deformation (since the base space P_1 is preserved) and so gives us ε on $S_x = \Gamma(L(1))$. This completes the definition of the metric, which turns out to be self-dual.

Finally, a few words about *gravitational instantons*. How one defines these depends on what one wants to use them for, but let us say that they are 4-dimensional Riemannian manifolds which

(a) are complete

(b) have vanishing Ricci tensor

(c) are either compact or (in some suitable sense) asymptotically flat.

Some of these spaces will be self-dual and will therefore correspond to deformations of twistor space satisfying some extra conditions. These extra conditions are not, as yet, clearly understood and a great deal of work remains to be done on the whole subject of gravitational instantons.

What I would like to mention here are some functionals which are closely analogous to the A_q of Theorem 4.2. Recall that L denotes the line bundle over $\hat{X}$ which is the lift of L(1) over P_1. Let q be a positive integer and look for holomorphic sections of L^q. Since $\Gamma(L(q)) \cong C^{q+1}$, we know that

$$\dim \Gamma(L^q) \geq q + 1 \tag{5.6}$$

(i.e. sections of L(q) over P_1 can be lifted to $\hat{X}$). A section of L^q over $\hat{X}$ corresponds, in the self-dual space-time X, to a solution ϕ of the *twistor equation*. This means that $\phi \in S^*(q)$ on X, with $(\tau\circ\nabla)\,\phi = 0$ [cf. Equation (2.2), where the twistor operator $\tau\circ\nabla$ is defined for flat space-time; the definition for curved space-time is the same].

Now in general the equality sign in (5.6) holds: all sections of L^q come from the base P_1 (equivalently, all solutions of the twistor equation are covariantly constant). Let G_q be the set of self-dual metrics on X admitting a *non-constant* solution of the twistor equation (i.e. so that $\dim \Gamma(L^q) > q + 1$). Then we might *conjecture* that all the elements of G_q arise as the images of functionals A_q, similar to those of Theorem 4.2, mapping solutions of a *linear* equation into the space of all self-dual metrics. This conjecture is true for q = 1,2: see [11] for the details of some of these functionals. For q > 2 not much is known. It is also not yet known whether there is an analogue of Theorem 4.3, i.e. whether all self-dual gravitational instantons can be generated by such functionals.

REFERENCES

[1] M.F. Atiyah, N.J. Hitchin, V.G. Drinfeld and Yu. I. Manin, Phys. Letters 65A (1978), 185-187.

[2] M.F. Atiyah, N.J. Hitchin and I.M. Singer, Self-duality in four-dimensional Riemannian geometry (preprint, Oxford, 1978; to appear in Proc. Roy. Soc.).

[3] M.F. Atiyah and R.S. Ward, Comm. Math. Phys. 55 (1977), 117-124.

[4] E.F. Corrigan, D.B. Fairlie, R.G. Yates and P. Goddard, Comm. Math. Phys. 58 (1978), 223-240.

[5] W.D. Curtis, D.E. Lerner and F.R. Miller, Some remarks on the non-linear graviton (to appear in Gen. Rel. Grav.)

[6] R.O. Hansen, E.T. Newman, R. Penrose and K.P. Tod, The metric and curvature properties of H-space (preprint, Pittsburgh, 1978; to appear in Proc. Roy. Soc.).

[7] R. Hartshorne, Comm. Math. Phys. 59 (1978), 1-15.

[8] S.W. Hawking, Phys. Letters 60A (1977), 81-83.

[9] R. Penrose, J. Math. Phys. 8 (1967), 345-366.

[10] R. Penrose, Gen. Rel. Grav. 7 (1976), 31-52.

[11] R.S. Ward, A class of self-dual solutions of Einstein's equations (preprint, Oxford, 1978; to appear in Proc. Roy. Soc.).

R.S. Ward
Merton College
Oxford
England

R Hartshorne

Algebraic vector bundles on projective spaces, with applications to the Yang-Mills equation

My purpose in these lectures is to give some idea of recent work in algebraic geometry concerning algebraic vector bundles on projective spaces, and how it is related to the solutions of the Yang-Mills equation. Since I have recently written three other articles on this subject, I will confine myself here to a general account of the most striking success in this area, namely how methods of algebraic geometry have led to a "complete" solution in terms of linear algebra of the problem of classification of instantons. This is only one of several recent instances of algebraic geometry appearing in connection with some of the differential equations of mathematical physics: two others are Hitchin's work on gravitational instantons (reported by Penrose at this conference) in which the rational double points of algebraic surfaces play a central rôle, and the work of Kričever, Mumford, and van Moerbeke relating certain line bundles on an algebraic curve to the solutions of the Korteweg-DeVries equation [10].

For further details I will refer to the survey article [7] about the connection between vector bundles and instantons, the detailed account [8] of the theory of rank 2 vector bundles on $\mathbb{P}^3$, and the list of open problems [9], which also contains a long bibliography. For general background in algebraic geometry see [6].

Jackiw in his talks at this conference explained why physicists are interested in instantons. The natural world should best be described in terms of quantum field theory, but the equations which arise in that theory are too difficult to solve. The corresponding classical equations give a first approximation. Then by allowing time to become imaginary, one obtains solutions of the classical equations which are forbidden in real time, and which are thought to reflect certain quantum-mechanically allowed motions such as "tunnelling" between two zero-energy states of a physical system. This leads in particular to the search for solutions of the classical Yang-Mills equation with imaginary time, i.e., in Euclidean space $\mathbb{R}^4$. For purposes of computation, the most important solutions will be those with minimal action: these are the self-dual (or anti-self-dual) solutions, which are called instantons. They are indexed by an integer $k \geq 0$, the "instanton number," and for each k one expects a continuous family of solutions depending on auxiliary parameters such as "position" and "scale." The physicists were able to find some solutions for each k, but could see by a parameter count that they had not found them all.

Ward in his talks at this conference (see also [3]) explained how the general ideas of Penrose allow one to transform the study of instantons into a problem of complex analysis. First we compactify $\mathbb{R}^4$ to the 4-sphere S^4. On S^4 for each integer k there is just one principal SU(2)-bundle with second Chern class $c_2 = k$ (we will stick to the group SU(2) for simplicity, although completely analogous results hold for all other compact gauge groups.) A self-dual solution to the Yang-Mills equation on S^4 corresponds to a connection on this SU(2)-bundle whose curvature is self-dual.

Now the Penrose transform appears in this case as a mapping $\pi: \mathbb{P}^3_{\mathbb{C}} \longrightarrow S^4$ from complex projective 3-space to S^4. Each point of S^4 corresponds to a complex projective line $\mathbb{P}^1_{\mathbb{C}}$ in $\mathbb{P}^3_{\mathbb{C}}$. This complex projective space has a real structure σ on it for which there are no real points, but there are real lines, namely those of the form $\pi^{-1}(x)$, for $x \in S^4$.

Given an SU(2)-bundle on S^4, one can lift it by π to an SU(2)-bundle on $\mathbb{P}^3_{\mathbb{C}}$, which is associated to a certain $\mathbb{C}^2$-bundle E. A connection with self-dual curvature on the original SU(2)-bundle gives an almost-complex structure on E, and the self-dual condition is the integrability condition needed to make E a holomorphic rank 2 vector bundle on $\mathbb{P}^3_{\mathbb{C}}$. Since it comes from S^4, it has a real structure $\tilde{\sigma}$ with $\tilde{\sigma}^2 = -1$, and clearly the restriction of E to each real line $\pi^{-1}(x)$ is trivial.

The main result of Atiyah and Ward [3] is that this is an equivalent problem: to give an SU(2)-instanton on S^4 is equivalent to giving a rank 2 holomorphic vector bundle E on $\mathbb{P}^3_{\mathbb{C}}$ whose restriction to each real line $\pi^{-1}(x)$ is trivial, and which has a real structure $\tilde{\sigma}$ lying over the real structure on $\mathbb{P}^3_{\mathbb{C}}$. The instanton number k appears as the second Chern class $c_2(E)$, while $c_1(E)$ is necessarily 0.

Thus we have arrived at a problem in algebraic geometry. Part of it is purely holomorphic, namely the study of holomorphic vector bundles on $\mathbb{P}^3_{\mathbb{C}}$. An old theorem of Serre states that every holomorphic vector bundle (and more generally every holomorphic coherent sheaf) on a complex projective space has a unique algebraic structure, so that the theory of holomorphic vector bundles is equivalent to the theory of algebraic vector bundles. The other part of the problem, involving the restriction to the real lines and the real structure, can be phrased as a problem in algebraic geometry over $\mathbb{R}$. Our approach will be to study the problem over $\mathbb{C}$ first, and then tack on the real structure afterwards.

The first step is to study the classification of algebraic rank 2 vector bundles on $\mathbb{P}^3_{\mathbb{C}}$. This is a problem which has concerned algebraic geometers for some time, in particular Horrocks, Barth, Maruyama, and myself. There are two numerical invariants, the Chern classes c_1, c_2, which can take on all integer values satisfying $c_1c_2 \equiv 0 \pmod 2$. Having fixed c_1, c_2, there are infinitely many families, depending on larger and larger parameter spaces, of bundles with the given Chern classes. The parameter spaces are not well-behaved topologically (e.g. they may not be Hausdorff spaces). So Mumford introduced the concept of a stable vector bundle. The stable bundles with given c_1, c_2 are parameterized by a finite union of quasi-projective varieties, which we call their moduli space. Also it happens that the bundles arising from instantons are all stable, so that we have good moduli spaces for them.

Fixing $c_1 = 0$, $c_2 > 0$ for simplicity (which includes in particular all the bundles coming from instantons), we can then speak of the moduli space $M(c_2)$ of stable algebraic rank 2 vector bundles on $\mathbb{P}^3_{\mathbb{C}}$ with those Chern classes. Unfortunately the structure of this space is rather complicated and not yet well understood. It may have several different irreducible components of different dimensions. It may not be connected. What is known is that every irreducible component must have dimension $\geq 8c_2 - 3$, and that there is at least one irreducible component of dimension exactly $8c_2 - 3$ which contains the bundles coming from the $(5c_2 + 4)$-dimensional family of instantons found by the physicists.

Here we have been talking about the complex dimension of the complex analytic space $M(c_2)$. Because of the reality conditions which are imposed on the bundles corresponding to instantons, the parameter space $M'(k)$ of instantons with instanton number k will be a union of certain connected components of the real part of $M(k)$. Thus it will be a real analytic space whose real dimension is equal to the complex dimension of $M(k)$. And indeed it is known that the moduli space $M'(k)$ of instantons is a real-analytic manifold of real dimension $8k - 3$ (this was proved by Atiyah, Hitchin, and

Singer [2] using methods of differential geometry). However, it is not yet known (except for $k = 1, 2$) whether $M'(k)$ is connected.

There are two principal methods for studying rank 2 vector bundles on $\mathbb{P}^3$. One is via curves, the other via monads.

The curve method goes back to Serre (1960), was used by Horrocks (1968), and was apparently rediscovered independently by Barth and Van de Ven (1974) and Grauert and Mülich (1975). It is the main technique used in [8]. The idea is this. Tensor the bundle E by a multiple of the Hopf line bundle $\mathcal{O}(1)$ so that the new bundle $E(n)$ has plenty of global sections. If s is a sufficiently general section, then the set of points in $\mathbb{P}^3$ where s becomes zero will be a curve Y in $\mathbb{P}^3$. Conversely given a curve Y together with certain extra data I won't state explicitly here, one can recover the bundle E.

Thus the study of vector bundles is reduced to the study of curves in $\mathbb{P}^3$. This is another difficult subject, but a lot is known, and it has a venerable history going back into the 19th century. This method leads easily to many examples of bundles. A first series of examples, for every $k > 0$, can be constructed by taking the curve Y to be a disjoint union of $k + 1$ lines in $\mathbb{P}^3$. These examples contain already all the bundles corresponding to instantons found by the physicists. A second series of examples for every $k > 0$ can be constructed by taking Y to be an elliptic curve of degree $k + 4$. For $k \geq 3$ these are new, not contained in the first series. There are other series after that, corresponding to curves of higher genus, but they cannot be described so explicitly because we do not know the corresponding curves so well.

The method of curves also lends itself well to the explicit description of bundles with small values of c_2. Thus for example all stable rank 2 bundles on $\mathbb{P}^3$ with $c_1 = 0$, and $c_2 = 1, 2$ have been classified, the corresponding moduli space described, and the moduli space of instantons with $k = 1, 2$ also described explicitly.

The other main method of studying bundles on $\mathbb{P}^3$ is by monads. According to Webster's International Dictionary, a monad is a unit or atom in Greek philosophy. In the philosophy of Leibnitz it is "an individual elementary being, psychical or spiritual in nature, but constituting the underlying reality of the physical as well." The word monad was used by Horrocks (1964) to denote a sequence

$$F' \xrightarrow{\alpha} F \xrightarrow{\beta} F''$$

of vector bundles and morphisms of bundles such that $\beta\alpha = 0$, α is injective, and β is surjective. The bundle $\ker\beta/\mathrm{im}\alpha$ is called the homology of the monad. Horrocks showed, in a letter to Mumford (1971), that every vector bundle E on $\mathbb{P}^3$ can be obtained as the homology of a monad in which each of the bundles F', F, F'' is a direct sum of line bundles. Thus the monad is a kind of "two-sided resolution" of the bundle E. Although a monad is a more complicated object than a single bundle E, it is easier to describe explicitly, because the individual bundles F', F, F'' are simple, and the maps α, β can be represented by matrices. Thus in principle the use of monads reduces the study of vector bundles to linear algebra.

Barth observed for the case of stable rank 2 bundles E on $\mathbb{P}^3$ with $c_1 = 0$, $c_2 = k$ that if one imposes the extra condition $H^1(E(-2)) = 0$, then the corresponding monad takes on the special form

$$\mathcal{O}(-1)^k \xrightarrow{\alpha} \mathcal{O}^{2k+2} \xrightarrow{\beta} \mathcal{O}(1)^k.$$

To specify such a monad, one has only to give α and β, which are $(2k+2) \times k$ and $k \times (2k+2)$ matrices of linear forms in the homogeneous coordinates z_0, z_1, z_2, z_3 of the projective space. Of course the matrices α, β are subject to the conditions that rank α = rank β = k at every point, and that $\beta\alpha = 0$. Barth showed further that this monad was self-dual, so that β is the transpose of α with respect to an alternating bilinear form J on $\mathcal{O}^{2k+2}$.

The extra cohomology condition $H^1(E(-2)) = 0$ remained somewhat mysterious until Atiyah and Hitchin, and independently Drinfeld and Manin, showed that this condition holds for all those bundles coming from instantons. The proof is by differential geometry on S^4. To quote their paper [1], it is because the linear differential equation $(\Delta + \frac{1}{6}R)u = 0$ has no global nonzero solutions, where u is a section of the SU(2)-bundle on S^4 corresponding to E, Δ is the covariant Laplacian of that bundle, and $R > 0$ is the scalar curvature of S^4. Putting these results together shows that every bundle coming from an instanton can be represented by a monad of the above special form.

To clarify this situation further, I would like to quote the form of Horrocks's monad construction used by Drinfeld and Manin (and explained very clearly in their paper [5]), which shows that the cohomological condition $H^1(E(-2)) = 0$ mentioned above is essentially equivalent to the possibility of representing E by a special monad. It is this. There is an equivalence of categories between

(i) the category of coherent sheaves F on $\mathbb{P}^3$ satisfying

$$H^0(F(m)) = 0 \quad \text{for } m \le -1$$
$$H^1(F(m)) = 0 \quad \text{for } m \le -2$$
$$H^2(F(m)) = 0 \quad \text{for } m \ge -2$$
$$H^3(F(m)) = 0 \quad \text{for } m \ge -3$$

and

(ii) special monads of the form

$$A \otimes \mathcal{O}(-1) \xrightarrow{\alpha} B \otimes \mathcal{O} \xrightarrow{\beta} C \otimes \mathcal{O}(1)$$

where A, B, C are complex vector spaces.

In the case of a bundle E of rank 2, the condition on H^0 is implied by stability, the condition on H^1 is implied by $H^1(E(-2)) = 0$, and the conditions on H^2 and H^3 follow by Serre duality.

The important point about this theorem is that it gives an equivalence of categories. This means that if the bundle E has any additional structure, then this additional structure is automatically inherited by the corresponding monad. In particular, if E has rank 2, then it carries an alternating bilinear form arising from the natural map $E \otimes E \longrightarrow \Lambda^2 E \cong \mathcal{O}$, so the corresponding monad also carries this form. If E comes from an instanton, then it has a real structure $\tilde{\sigma}$, and this also carries over to the monad.

This achieves the reduction of the problem of classification of instantons to a problem in linear algebra. Each SU(2)-instanton with instanton number k corresponds to a stable rank 2 vector bundle on $\mathbb{P}^3_{\mathbb{C}}$ which can be represented by a special monad as above. The monad is self-dual, and the reality conditions carry over. To describe the monad, we have only to give the map α, and list the conditions it must satisfy. The map α is given by a $(2k + 2) \times k$ matrix $A(z)$ of linear forms in the homogeneous coordinates z_0, z_1, z_2, z_3 of the projective space. It is subject to the conditions

(1) $A(z)^t J A(z) = 0$ for all z

(2) $\operatorname{rank} A(z) = k$ for all $z \neq 0$

(3) $J\,\overline{A(z)} = A(\sigma z)$

where J is the $(2k + 2) \times (2k + 2)$ skew symmetric matrix $\begin{pmatrix} 0 & I \\ -I & 0 \end{pmatrix}$ and $\sigma(z_0,z_1,z_2,z_3) = (-\bar{z}_1,\bar{z}_0,-\bar{z}_3,\bar{z}_2)$.

Condition (1) expresses the relation $\beta\alpha = 0$ when β is taken to be the transpose of α via self-duality of the monad; (2) is the rank condition which makes α injective and β surjective; (3) is the reality condition.

There is ambiguity in the choice of bases for the bundles on the left and in the middle of the monad. Thus two matrices A(z) and A'(z) give the same instanton if and only if

$$A'(z) = S \cdot A(z) \cdot T$$

where $S \in Sp(k + 1)$ and $T \in GL(k,\mathbb{R})$. So the moduli space M'(k) of instantons can be obtained as the space of all such possible matrices A(z) modulo the action of the group $Sp(k + 1) \times GL(k,\mathbb{R})$.

Some explicit calculations of the moduli space have been made from this point of view by Rawnsley, Jones, and by Christ in his talk at this conference (see [4]). While this matrix formulation has been hailed by some as a "complete solution" of the problem of classifying instantons, there are still many questions left unanswered. The calculations quickly become very complicated, so that for example one still does not have an answer to the (deceptively) simple question whether the moduli space M'(k) of instantons is connected for $k \geq 3$.

References

1 M. F. Atiyah, N. J. Hitchin, V. G. Drinfeld, Ju. I. Manin, Construction of instantons, Physics Letters 65A (1978) 185-187.

2 M. F. Atiyah, N. J. Hitchin, I. M. Singer, Deformations of instantons, Proc. Nat. Acad. Sci. USA, 74 (1977) 2662-2663.

3 M. F. Atiyah, R. S. Ward, Instantons and algebraic geometry, Comm. Math. Phys. 55 (1977) 117-124.

4 N. H. Christ, E. J. Weinberg, N. K. Stanton, General self-dual Yang-Mills solutions (preprint)

5 V. G. Drinfeld, Ju. I. Manin, Instantons and sheaves on $\mathbb{CP}^3$ (preprint)

6 R. Hartshorne, Algebraic Geometry, Graduate Texts in Math 52, Springer-Verlag, New York (1977) xvi + 496 pp.

7 R. Hartshorne, Stable vector bundles and instantons, Comm. Math. Phys. 59 (1978) 1-15.

8 R. Hartshorne, Stable vector bundles of rank 2 on $\mathbb{P}^3$, Math. Ann. (to appear)

9 R. Hartshorne, Algebraic vector bundles on projective spaces: a problem list, Topology (to appear).

10 D. Mumford, An algebro-geometrical construction of commuting operators and of solutions to the Toda lattice equation, Korteweg-De Vries equation and related non-linear equations, Kyoto Conference (to appear)

ROBIN HARSHORNE
Department of Mathematics
University of California
Berkeley, CA 94720

N H Christ
Self-dual Yang-Mills solutions

Let us consider the application of the Horrocks-Barth construction to the problem of finding self-dual Euclidean Yang-Mills solutions, recently developed by Atiyah, Hitchin, Drinfeld and Manin [1]. In this talk I should like to show explicitly how the construction of these authors can be reduced to the problem of solving a non-linear equation for a matrix of quaternions. As we will see, this quaternionic formulation has a very simple and direct connection with the original, physical gauge theory and yields, for example, all SU(2) solutions with Pontryagin index k=3 and Sp(n) solutions with $k \leq n$.

The construction is simplest for the symplectic groups Sp(n) (where Sp(1)=SU(2)). One begins with the monad

$$W^k \xrightarrow{M^\alpha z_\alpha} V^{2k+2n} \xrightarrow{(M^\alpha z_\alpha)^*} W^* \qquad (1)$$

where W^k and V^{2k+2n} are complex vector spaces of dimension k and 2k+2n respectively while the sum $M^\alpha z_\alpha$ is an injection of W^k into V^{2k+2n} which depends linearly on the four homogeneous coordinates $\{z\}_\alpha\ _{1 \leq \alpha \leq 4}$ of $\mathbb{CP}_3$. Furthermore V^{2k+2n} is identified with its dual by introducing a symplectic bilinear form Ω^{k+n} defined on V^{2k+2n}-thus allowing the dual mapping, $(M^\alpha z_\alpha)^*$, to act on V^{2k+2n}. The fiber of the resulting rank two vector bundle over the point z in $\mathbb{CP}_3$ is given by

$$(E_2)_z = \ker\, (M^\alpha z_\alpha)^* / \mathrm{im}(M^\alpha z_\alpha). \qquad (2)$$

In order to obtain a bundle corresponding to a symplectic gauge group we must introduce conjugate linear involutions σ_W and σ_V on W^k and V^{2k+2n} obeying

$$\sigma_W^2 = 1, \quad \sigma_V^2 = -1 \tag{3}$$

and demand

$$\sigma_V[M^\alpha z_\alpha] = M^\alpha \sigma(z)_\alpha \, \sigma_W \tag{4}$$

where σ is the mapping on $\mathbb{CP}_3$

$$\sigma(z_1,z_2,z_3,z_4) = (\bar{z}_2,-\bar{z}_1,\bar{z}_4,-\bar{z}_3). \tag{5}$$

We can choose a basis for W^k and V^{2k+2n} with respect to which σ_W is simply complex conjugation while σ_V is complex conjugation followed by transformation with the $2(k+n) \times 2(k+n)$ matrix

$$\Omega^{k+n} = \begin{pmatrix} 0 & 1 & 0 & 0 & & & \\ -1 & 0 & 0 & 0 & & & \\ 0 & 0 & 0 & 1 & & & \\ 0 & 0 & -1 & 0 & & & \\ & & & & \ddots & & \\ & & & & & 0 & 1 \\ & & & & & -1 & 0 \end{pmatrix}. \tag{6}$$

Thus our bundle has a conjugation σ_E which lies over the operator σ of Eq. (5) on $\mathbb{CP}_3$ and a symplectic, bilinear form induced by the form Ω^{k+n}.

The connection between the conjugation σ_E and our search for symplectic gauge fields can be seen by referring to Ward's construction [2]. Recall that a symplectic matrix S in Sp(n) is a $2n \times 2n$ unitary matrix

$$S^\dagger S = I \tag{7}$$

which preserves the symplectic form Ω^n:

$$S^t \Omega^n \, S = \Omega^n. \tag{8}$$

Combining these equations one obtains,

$$\Omega^n \bar{S} = S\Omega^n \tag{9}$$

where $\bar{S}$ is the complex conjugate of the matrix S. Thus for real points $x \in \mathbb{C}^4$ a symplectic connection matrix $A_\mu(x)$ will obey Eq. (9) and the transition functions obtained in the Ward construction,

$$G_{ij}(z) = P\{\exp \int_{x^i(z)}^{x^j(z)} A_\mu(x)dx_\mu\} \tag{10}$$

will be compatible with a conjugation σ_E similar to that induced on E_2 by Eq. (4):

$$G_{ij}(\sigma(z))\Omega^n = \Omega^n \bar{G}_{ij}(z). \tag{11}$$

The right hand side of Eq. (10) is the usual path-ordered integral. The points $x^i(z)$ and $x^j(z)$ lie on the anti-self dual plane determined by the point z in $\mathbb{CP}_3$,

$$\begin{pmatrix} z_3 \\ z_4 \end{pmatrix} = \begin{pmatrix} x_0 - i\, x_3 & -x_2 - i\, x_1 \\ +x_2 - i\, x_1 & x_0 + i\, x_3 \end{pmatrix} \begin{pmatrix} z_1 \\ z_2 \end{pmatrix} \tag{12}$$

(where $\{x_\mu\}_{0 \le \mu \le 3}$ are the four components of the point x), and are chosen to obey

$$\bar{x}(z) = x(\sigma(z)). \tag{13}$$

Finally, the condition (8) implies in a more straight forward way the existence of a symplectic bilinear form on the resulting bundle:

$$G_{ij}(z)^t \Omega^n G_{ij}(z) = \Omega^n. \tag{14}$$

The preceeding arguments can be easily transcribed to the orthogonal groups O(n), in which case Eq. (8) and the reality condition (11) apply without the matrix Ω. Consequently in the Atiyah, Hitchen, Drinfeld, Manin construction $\sigma_V^2 = +1$ so that one must choose $\sigma_W^2 = -1$ if Eq. (4) is to be consistent with the property $\sigma^2 = -1$, implied by Eq. (5). The unitary groups, U(n), are somewhat different since the fundamental representation of U(n), n>2, is inequivalent to the complex conjugate representation. Instead, for U(n), n>2, one can use Eq. (7) and introduce a conjugate linear mapping which relates the bundle E_2 with its dual E_2^*.

Let us now develop a simple representation for the mapping $M^\alpha z_\alpha$ of Eq. (1) and the implied requirement

$$\ker[(M^\alpha z_\alpha)^*] \supset \operatorname{im}[M^\alpha z_\alpha]. \tag{15}$$

If we introduce a spinor notation

$$(z_1, z_2, z_3, z_4) = (\omega_1, \omega_2, \pi_1, \pi_2) \tag{16}$$

so that Eq. (12) becomes

$$\omega_A = x_A^{\ B}\, \pi_B \ , \tag{17}$$

then $M^\alpha z_\alpha$ can be written as the matrix

$$B^A \pi_A - C^A \omega_A = (B^A - C^{A'} x_{A'}^{\ A}) \pi_A = M^A(x) \pi_A \tag{18}$$

Using the basis previously adopted, the reality condition (4) becomes

$$\Omega \bar{B}_A = B^A, \quad \Omega \bar{C}_A = C^A \tag{19}$$

where for a spinor ψ_A

$$\begin{pmatrix}\psi_1\\ \\ \psi_2\end{pmatrix}=\begin{pmatrix}\psi 2\\ \\ -\psi^1\end{pmatrix}. \tag{20}$$

Thus if $B^A_{2\ell+a-1,m}$ and $C^A_{2\ell+a-1,m}$ for fixed ℓ and m are viewed as 2×2 matrices with indicies a and A, they belong to a 2×2 representation of real quaternions:

$$1=\begin{pmatrix}1 & 0\\ 0 & 1\end{pmatrix} \qquad \hat{i}=\begin{pmatrix}0 & -i\\ -i & 0\end{pmatrix}$$

$$\hat{j}=\begin{pmatrix}0 & -1\\ 1 & 0\end{pmatrix} \qquad \hat{k}=\begin{pmatrix}-i & 0\\ 0 & i\end{pmatrix}. \tag{21}$$

Consequently we will view B,C, and $M(x) = B-Cx$ as $(n+k)\times k$ matrices of quaternions where x is the quaternion represented by the matrix in Eq. (12).

The requirement (15) can be written

$$(M^\alpha z_\alpha)^t \Omega\, M^\beta z_\beta = 0 \tag{22a}$$

or

$$0 = (M^\alpha)^\dagger \sigma(\bar{z})_\alpha M^\beta z_\beta = \bar{\pi}^A [M^\dagger(x)M(x)]_A{}^B \pi_B \tag{22b}$$

Where $M^\dagger(x)$ is the transpose of the matrix whose elements are the quaternionic conjugates of those in M(x):

$$(M(x)^\dagger)_{\ell m} = (M(x)_{m\ell})^\dagger \tag{23}$$

with $(1)^\dagger = 1$, $(\hat{i})^\dagger = -\hat{i}$, $(\hat{j})^\dagger = -\hat{j}$ and $(\hat{k})^\dagger = -\hat{k}$. The indicies A and B in Eq. (22b) again refer to the 2×2 matrix description, Eq. (21), of the quaternions. The validity of Eq. (22b) for all π requires that the matrix

$$R(x) = M^\dagger(x)M(x) \tag{24}$$

be real. Thus the problem of finding self-dual Sp(n) gauge fields has been reduced to that of finding a $(k+n)\times k$ matrix of quaternions $M(x)$ of the form $B - Cx$ such that the product $M^{\dagger}(x)M(x)$ is a real, non-singular matrix.

This matrix formulation has the added practical advantage that the gauge field, of physical interest, can be directly constructed in terms of it [3,4]. First define a $(k+n)\times n$ quaternionic matrix $N(x)$ obeying

$$N^{\dagger}(x)N(x) = I \tag{25a}$$

$$N^{\dagger}(x)M(x) = 0. \tag{25b}$$

Then the gauge field corresponding to the original bundle E_2 is given by

$$A_{\mu}(x) = N^{\dagger}(x)\partial_{\mu}N(x). \tag{26}$$

If the quaternionic elements of $A_{\mu}(x)$ are replaced by 2×2 matrices as in Eq. (21) and the result viewed as a $2n\times 2n$ matrix, one obtains a skew-hermitian gauge field obeying Eq. (8), i.e., a symplectic gauge field. It is easy to show directly that the field strength corresponding to $A_{\mu}(x)$ is self-dual. Finally, and perhaps most remarkably, the matrix $N(x)$ can be used to find a solution [3,4] to the equation

$$D_{\mu}D_{\mu}\Delta(x,y) = \delta^{4}(x-y), \tag{27}$$

where

$$D_{\mu} = \frac{\partial}{\partial x_{\mu}} + A_{\mu}(x). \tag{28}$$

The solution to Eq (27) is simply

$$\Delta(x,y) = \frac{1}{4\pi^{2}}\,\frac{N^{\dagger}(x)N(x)}{(x-y)^{2}}\,. \tag{29}$$

As a last topic we will discuss two cases in which this simple quaternonic formulation allows the explicit construction of new, self-dual solutions. We can choose a canonical form for the matrix C in which the first n rows vanish and the last k rows make up a k×k unit matrix. With this choice of C the condition that $M^{\dagger}(x)M(x)$ be real, requires that $B^{\dagger}B$ be real and that the last k rows of B make up a symmetric matrix. If we let $\{q_{ij}\}_{1\le i\le n,1\le j\le k}$ be the matrix formed by the first n rows of B and $\{b_{ij}\}_{1\le i,j\le k}$ that formed by the remaining k rows, these conditions are that $b_{ij} = b_{ji}$ and that

$$\sum_{\ell=1}^{n} (q_{\ell i})^{\dagger} q_{\ell j} + \sum_{\ell=1}^{k} (b_{\ell i})^{\dagger}(b_{\ell j}) \tag{30}$$

be real. Thus the q_{ij} represent 4nk real parameters while the symmetric b_{ij} represent 2k(k+1). The reality of the k×k matrix (30) imposes $\frac{3}{2}k(k-1)$ conditions leaving $4nk + \frac{1}{2}k(k+7)$ parameters. However, the self-dual solution (26) is not altered and our canonical form for C and the condition (30) preserved if we multiply M(x) on the right by a k×k orthogonal matrix 0 (k(k-1)/2 parameters) and on the left by a (n+k) × (n×k) matrix S made up of an n×n symplectic matrix G and 0^{-1}:

$$S = \begin{pmatrix} G & 0 \\ 0 & 0^{-1} \end{pmatrix} \tag{31}$$

(n(2n+1) parameters). Thus there remain 4(n+1)k-n(2n+1) meaningful parameters--the number given previously [5] for general Sp(n) self-dual solutions when $k\ge n$. When $k\le n$, the above counting must be modified since the n×n symplectic matrix G acts on only k×n independent columns of q_{ij}. Consequently q_{ij} will be unaffected by a Sp(n-k) subgroup of Sp(n), yielding an additional (n-k)(2n-2k+1) parameters for a total of $2k^2 + 3k$.

In fact, in the case $n \geq k$, a general solution to the nonlinear requirement that the expression (30) be real can be written down [6]. The elements $b_{ij} = b_{ji}$ are first chosen arbitrarily ($2k(k+1)$ parameters). Next the freedom implied by the Sp(n) transformations G in Eq. (31) is used to choose $q_{i1} = 0$ for $i > 1$ and q_{11} real and positive. Next, this symplectic freedom is exploited to choose $q_{i2} = 0$ for $i>2$ and q_{22} real and positive. The reality of Eq. (30) for $i=1$, $j=2$ then fixes the quaternionic part of q_{12}. This process can be continued inductively producing a matrix q_{ij} which vanishes for $i>j$, is real and positive for $i=j$, and has a determined quaternionic part for $i<j$. Thus the reality of the expression (30) is quaranteed and q_{ij} contains $k(k+1)/2$ real parameters for a total of $\frac{5}{2}k^2+\frac{3}{2}k$--consistent with out previous result since the O(k) redundancy has not been removed. A similar result can also be obtained for O(n), $4n \geq k$, and SU(n), $2n \geq k$.

For fixed n and general k it is not yet known how to write down the general solution. However, the 5k parameter solutions of 't Hooft can be easily obtained in this language. Choose the q_{1i} to be real and positive (the square root of the ith instanton scale) and b_{ij} to be diagonal (b_{ii} is the position of the ith instanton).

A second case [4] in which an explicit general form for B_{ij} can be found consistent with the reality of the quantity (30) is the case of Sp(1)=SU(2) and k=3. We first use the Sp(1) × O(3) symmetry of our formulation to choose q_{11} real and to diagonalize the real part of b_{ij}. The remaining components of b_{ij} are then treated as the 21 = 8k-3 physical parameters and the q_{1i}, $1 \leq i \leq 3$, fixed by the requirement that the quantity (30) be real. The result is

$$q_1 = \frac{(W_2W_3-W_3W_2)}{2\sqrt{re(W_1W_3W_2)}}$$

$$q_2 = q_1 \frac{re[W_3W_2(W_1W_3-W_3W_1)]}{2|W_2W_3-W_2W_2|^2} - \frac{1}{q_1} W_3 \tag{32}$$

$$q_3 = q_1 \frac{re[W_3W_2(W_1W_2-W_2W_1)]}{2|W_2W_3-W_3W_2|^2} + \frac{1}{q_1} W_2$$

where the quaternions W_k are defined by

$$W_k = \frac{1}{4} \varepsilon_{ijk} \sum_{\ell=1}^{n} b^\dagger_{\ell i} b_{\ell j} - b^\dagger_{\ell j} b_{\ell i} \tag{33}$$

In conclusion, it appears that this very explicit quaternionic description of the Horrocks, Barth-Atiyah, Hitchin, Drinfeld, Manin construction is a very powerful one, both allowing the construction of new solutions and making very accessible quantities of physical interest such as the gauge field $A_\mu(x)$ and the isospin ½ Green's function $\Delta(x,y)$.

References

1 V. G. Drinfeld and Yu. I. Manin, Funkcional Anal. i. Prilozen (to appear);
V. G. Drinfeld and Yu. I. Manin, Uspehi Mat. Nauk. (to appear);
M. F. Atiyah, N. J. Hitchin, V. G. Drinfeld and Yu. I. Manin, Phys. Lett., 65A, 285(1978).

2 R. S. Ward, Phys. Lett., 61A, 81(1977).

3 E. J. Corrigan, D. B. Fairlie, S. Templeton and P. Goddard, A. Green's Function for the General Self-Dual Gauge Field, preprint.

4 N. H. Christ, E. J. Weinberg and N. K. Stanton, Columbia University preprint, # CU-TP-119.

5 C. W. Bernard, N. H. Christ, A. H. Guth and E. J. Weinberg, Phys. Rev. D16, 2967(1977) and reference therein.

6 V. G. Drinfeld, and Yu. I. Manin, A Description of Instantons, preprint.

Acknowledgment

This research was supported in part by the United States Department of Energy.

N H CHRIST
Department of Physics
Columbia University
New York, New York 10027

R Penrose
On the twistor descriptions of massless fields

1. TWISTOR FUNCTIONS AND SHEAF COHOMOLOGY *

Recall [1,2] the following properties of a twistor function $f(Z^\alpha)$, to be used for generating a zero rest-mass field $\phi_{A'...L'}$, or $\phi_{A...K}$, by means of a contour integral,

$$\phi_{A'...L'} = \frac{1}{(2\pi i)^2} \oint \pi_{A'} \cdots \pi_{L'} \, f(Z^\alpha) d^2\pi,$$

$$\phi_{A...K} = \frac{1}{(2\pi i)^2} \oint \frac{\partial}{\partial \omega^A} \cdots \frac{\partial}{\partial \omega^K} \, f(Z^\alpha) d^2\pi:$$

(i) The function f is holomorphic on some domain $\mathcal{D}$ of twistor space <u>not</u> invariant under SU(2,2) (nor under the Poincaré group, nor the Lorentz group).

(ii) There is a "gauge" freedom G whereby

$$f \mapsto f + h^- + h^+,$$

where $h^\pm$ is holomorphic on some extended domain $\mathcal{D}^\pm (\supset \mathcal{D})$ of twistor space in which the contour γ can be deformed to a point (to the "left" in $\mathcal{D}^-$ and to the "right" in $\mathcal{D}^+$)

*Taken (with minor corrections) from Twistor Newsletter 2 (10 June, 1976).

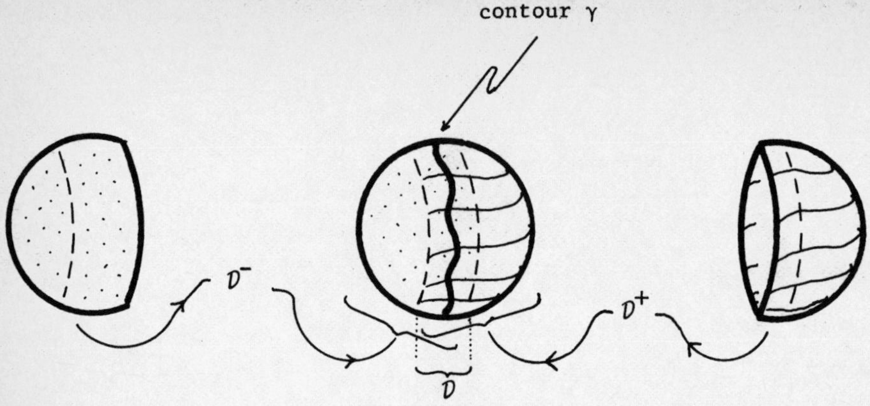

Domain $\mathcal{D}^- \cup \mathcal{D}^+$ (e.g. $\mathbb{T}^+$) symbolically represented as S^2.

(iii) G depends on the location of γ (and γ is not invariant under SU(2,2)--nor under the Poincaré group nor the Lorentz group--so G is not invariant either).

(iv) By invoking G, then moving γ, then invoking a new G, moving γ again,

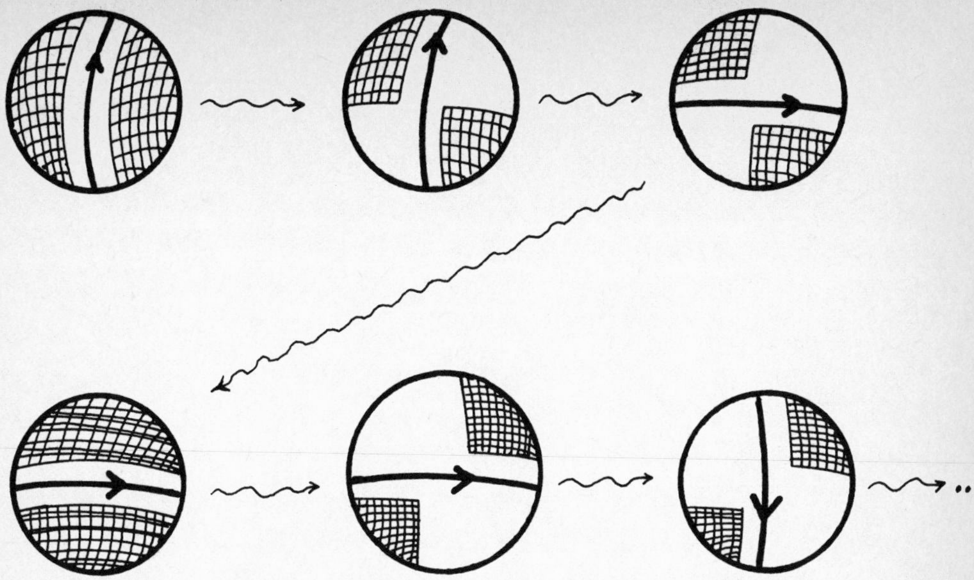

we can obtain a whole family of equivalent twistor functions, all giving the same field, the entire family being invariant under SU(2,2).

(v) This seems a little nebulous, and, for example, how do we add two such families (to give $\phi_{A'\ldots L'} + \psi_{A'\ldots L'}$), etc. etc?

A new viewpoint concerning twistor functions has been gradually emerging which makes good mathematical sense of all this: a twistor function is really to be viewed as a representative function (or cocycle) defining an element of a sheaf cohomology group. Now, the twistor theorist, when attacked by a purist for shoddiness in the domains, can counter-attack armed with his sheaf!

Thumbnail sketch of (relevant) sheaf cohomology theory

First let us recall how ordinary (Čech) cohomology works. Let X be a space (a Hausdorff paracompact topological space, say). Cover X with a locally finite system of open sets $\mathcal{U}_i$. We define a cochain (with respect to this covering) with coefficients in an additive abelian group $\mathbb{G}$ (say, the integers $\mathbb{Z}$, the reals $\mathbb{R}$, or the complex field $\mathbb{C}$) in terms of a collection of elements $f_i, f_{ij}, f_{ijk}, \ldots \in \mathbb{G}$, assigned to the various $\mathcal{U}_i$ and their non-empty intersections: f_i assigned to $\mathcal{U}_i$; f_{ij} assigned to $\mathcal{U}_i \cap \mathcal{U}_j$; f_{ijk} assigned to $\mathcal{U}_i \cap \mathcal{U}_j \cap \mathcal{U}_k \ldots$, and $f_{ij} = -f_{ji}$, $f_{ijk} = =f_{jik} = f_{jki} = \ldots, \ldots$, i.e. $f_{i\ldots\ell} = f_{[i\ldots\ell]}$. Then

$$0\text{-cochain } \alpha = (f_1, f_2, f_3, \ldots)$$

$$1\text{-cochain } \beta = (f_{12}, f_{23}, f_{13}, \ldots)$$

$$2\text{-cochain } \gamma = (f_{123}, f_{124}, \ldots)$$

(where $\mathcal{U}_1 \cap \mathcal{U}_2$, $\mathcal{U}_2 \cap \mathcal{U}_3$, $\mathcal{U}_1 \cap \mathcal{U}_3$, ... ,$\mathcal{U}_1 \cap \mathcal{U}_2 \cap \mathcal{U}_3$, $\mathcal{U}_1 \cap \mathcal{U}_2 \cap \mathcal{U}_4$,... are the non-empty intersections of $\mathcal{U}_i$'s).

Define <u>coboundary</u> <u>operator</u> δ as follows:

$$\delta\alpha = (\underset{\text{"}f_{12}\text{"}}{f_2 - f_1},\ \underset{\text{"}f_{23}\text{"}}{f_3 - f_2},\ \underset{\text{"}f_{13}\text{"}}{f_3 - f_1}, \ldots)$$

$$\delta\beta = (\underset{\text{"}f_{123}\text{"}}{f_{12} - f_{13} + f_{23}},\ \underset{\text{"}f_{124}\text{"}}{f_{12} - f_{14} + f_{24}}, \ldots)$$

etc.

(where, again, $\mathcal{U}_1 \cap \mathcal{U}_2$,... are the non-empty intersections). Then we have $\delta^2 = 0$. We call γ a <u>cocycle</u> if $\delta\gamma = 0$; we call γ a <u>coboundary</u> if $\gamma = \delta\beta$ for some β. Define the <u>p^{th}</u> <u>cohomology</u> <u>group</u> by:

$$H^p_{\{\mathcal{U}_i\}}(X, \mathbb{G}) = \begin{pmatrix}\text{additive group}\\ \text{of p-cocycles}\end{pmatrix} \Big/ \begin{pmatrix}\text{additive group}\\ \text{of p-coboundaries}\end{pmatrix}.$$

<u>Note</u>: $H^p_{\{\mathcal{U}_i\}}(X, \mathbb{G})$, as defined, depends on the covering $\{\mathcal{U}_i\}$. What we should do, to define $H^p(X, \mathbb{G})$, is to take the appropriate "limit" of all these

$H^p_{\{U_i\}}(X,\mathbb{G})$ for finer and finer coverings $\{U_i\}$ of X. However (for X suitably non-pathological) we can always settle on a particular "sufficiently fine" covering $\{U_i\}$ where, in effect, there is no "relevant topology" left on each U_i or intersections thereof (i.e. all the $H^p(U_i \cap \ldots \cap U_k, G)$ vanish for all $p > 0$--although, as it stands, this is somewhat unhelpful because a direct limit is then already involved in the definition of "sufficiently fine"). Then <u>this</u> $H^p_{\{U_i\}}(X,\mathbb{G}) = H^p(X,\mathbb{G})$. I shall henceforth assume that such a "sufficiently" fine covering has been taken, and that it is countable and locally finite.

Now what does this definition have to do with the familiar "dual" relation to ordinary homology $H_p(X,\mathbb{G})$? How does γ assign values (elements of $\mathbb{G}$) in a linear fashion to p-cycles in X, where γ is some element of $H^p(X,\mathbb{G})$? Intuitively:

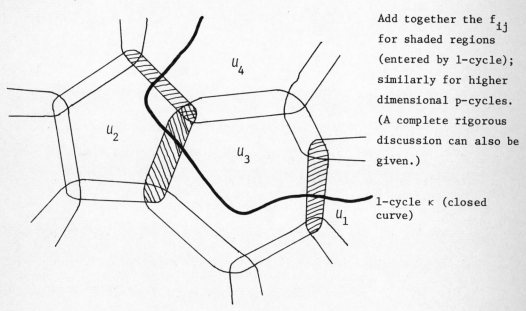

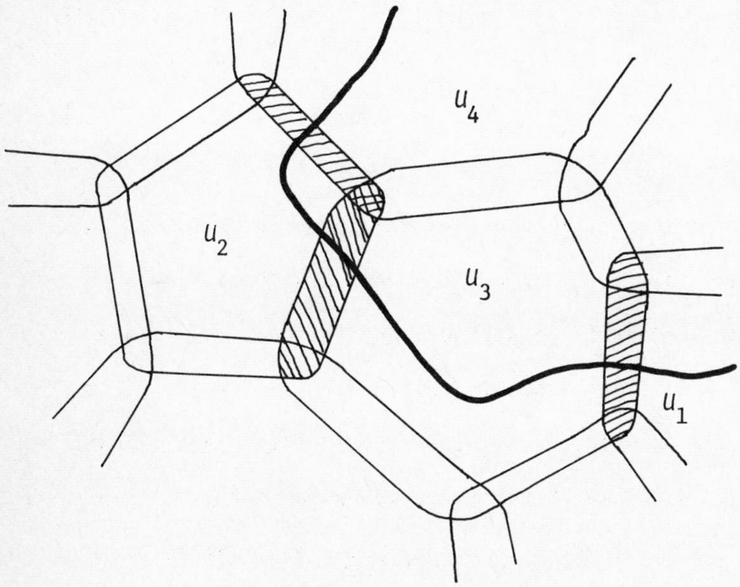

Add together the f_{ij} for shaded regions (entered by 1-cycle); similarly for higher dimensional p-cycles. (A complete rigorous discussion can also be given.)

1-cycle κ (closed curve)

If γ is defined by $(f_{12}, f_{23}, \ldots)$, then $\gamma(\kappa) = f_{42} + f_{23} + f_{31} + \ldots$ (correctly signed) f's on regions entered by κ.

What about sheaf cohomology though? It's really a rather natural generalization of the above. But first we must rephrase the concept of a cochain slightly. Rather than thinking of f_i as simply an element of $\mathbb{G}$ "assigned" to $\mathcal{U}_i$, and f_{ij} "assigned" to $\mathcal{U}_{ij}$, etc., we think of f_i as a function defined on $\mathcal{U}_i$ which happens, in the above, to take this constant value "f_i" $\in \mathbb{G}$, and we think of f_{ij} as a constant function on $\mathcal{U}_i \cap \mathcal{U}_j$ with values in $\mathbb{G}$, etc. (This has the incidental advantage that the requirement $\mathcal{U}_i \cap \mathcal{U}_j \neq \emptyset$ is now unnecessary.) This is still ordinary (Čech) cohomology. But the generalization to sheaf cohomology is now easily made: the functions f_i, f_{ij}, $f_{ijk}, \ldots$ are now not required to be constant. (In fact, we could even allow that the additive group $\mathbb{G}$, in which the values of the f's reside, may vary from point to point in X, or, indeed, the f's need not really be "functions" in the ordinary sense at all--but such situations will not be considered here.) We may require that the f's be restricted in some way--in particular, for the purposes of the present applications we shall often require the f's to be holomorphic (with, here, $\mathbb{G} = \mathbb{C}$) and X a complex manifold--or, we may consider other related classes of functions.

So, what's a sheaf? Actually, I shan't even bother with a formal definition (which can be found in, for example, the references [3,4,5]). The essential point is that a sheaf is so defined that the Čech cohomology works just as well as before. In fact, a sheaf $\mathcal{S}$ defines an additive group $\mathbb{G}_{\mathcal{U}}$ for each open set $\mathcal{U} \subset X$. For example, $\mathbb{G}_{\mathcal{U}}$ might be the additive group of all holomorphic functions on $\mathcal{U}$ (taking X to be a complex manifold). In this case we get the sheaf, denoted $\mathcal{O}$, of germs of holomorphic functions on X. Slightly more generally, we might consider "twisted" holomorphic functions, i.e. functions whose values are not just ordinary complex numbers, but taken in some complex line bundle over X (think of "spin-weighted" functions, for example). An important example of such a twisted function would arise if X were taken to be projective twistor space $\mathbb{P}\mathbb{T}$ (or a suitable portion thereof) and the functions considered were to be homogeneous (and holomorphic) of some fixed degree n in the twistor variable. For each open set $\mathcal{U} \subset X$ we take $\mathbb{G}_{\mathcal{U}}$ to consist of all such twisted functions on $\mathcal{U}$, and the resulting sheaf denoted $\mathcal{O}(n)$, is called the "sheaf of germs of holomorphic functions twisted by n" (on X). More generally we might consider functions whose values lie in some vector bundle B over X (e.g. we might consider tensor fields on X) and $\mathbb{G}_{\mathcal{U}}$ would consist of the cross-sections of the portion of B lying above $\mathcal{U}$:

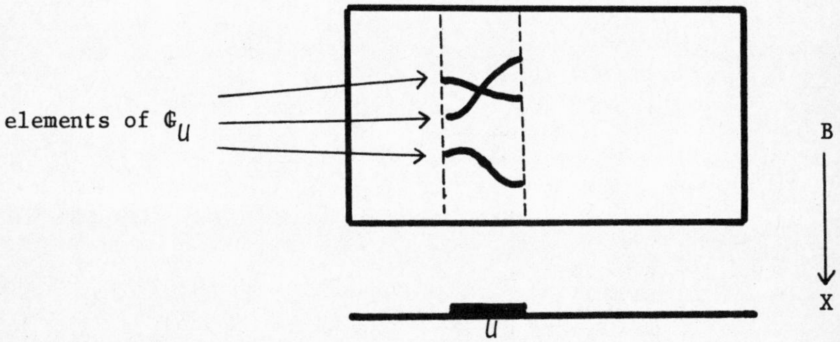

Cochains are defined as before (with $f_i \in \mathbb{G}_{\mathcal{U}_i}$, $f_{ij} \in \mathbb{G}_{\mathcal{U}_i \cap \mathcal{U}_j}, \ldots$) and the coboundary operator δ, just as before. Then we obtain the pth cohomology group of X, with coefficients in the sheaf S, as

$$H^p(X,S) = \left(\begin{matrix}\text{p-cochains with}\\ \text{coefficients in } S\end{matrix}\right) \Big/ \left(\begin{matrix}\text{p-coboundaries with}\\ \text{coefficients in } S\end{matrix}\right).$$

As before, we would need to take the appropriate limit for finer and finer coverings $\{\mathcal{U}_i\}$ of X, but we can settle on one "sufficiently fine" covering if desired. Provided $\mathcal{S}$ is what is called a coherent analytic sheaf (and we are interested primarily in such sheaves--locally defined by n holomorphic functions factored out, if desired, by a set of s holomorphic relations), then "sufficiently fine" can be taken to mean that each of $\mathcal{U}_i$, $\mathcal{U}_i \cap \mathcal{U}_j$, $\mathcal{U}_i \cap \mathcal{U}_j \cap \mathcal{U}_k, \ldots$ is a Stein manifold [4] (and it is sufficient just to specify that each $\mathcal{U}_i$ is Stein). In effect, a Stein manifold is a holomorphically convex open subset of $\mathbb{C}^n$ (or a domain of holomorphy). If X is Stein and $\mathcal{S}$ coherent, then $H^p(X,\mathcal{S}) = 0$, if $p > 0$. Note: $\mathcal{O}$ and $\mathcal{O}(n)$ are coherent.

Twistor functions as elements of $H^1(X,\mathcal{O}(n))$

Let X be some suitable portion of projective twistor space $\mathbb{PT}$, say some neighborhood of a line in $\mathbb{PT}$ (corresponding to some neighborhood of a point in Minkowski space), or say $\mathbb{PT}^+$, or $\overline{\mathbb{PT}^+}$. Suppose we can cover X with two sets $\mathcal{U}_1, \mathcal{U}_2$ (each open in X) such that every projective line L in X meets $\mathcal{U}_1 \cap \mathcal{U}_2$ in an annular region and where $\mathcal{U}_1 \cap \mathcal{U}_2$ corresponds to the domain of definition of some twistor function $f(Z^\alpha)$, homogeneous of degree n in the twistor Z^α.

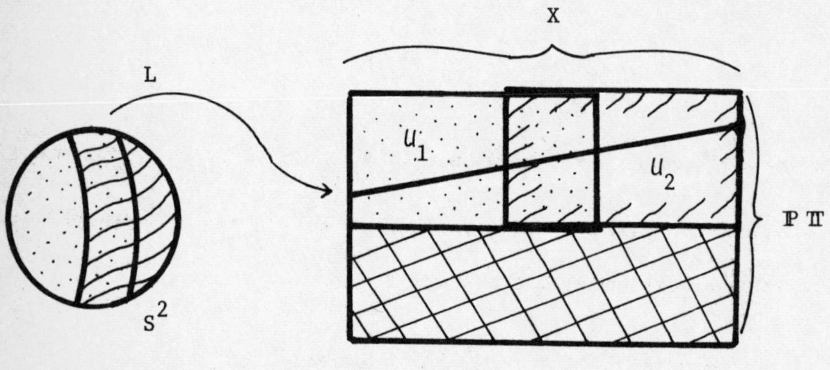

Then $f = f_{12}$ is a twisted function on $\mathcal{U}_1 \cap \mathcal{U}_2$. There are no other $\mathcal{U}_i \cap \mathcal{U}_j$'s, so f_{12} by itself defines a 1-cochain β, with coefficients in $\mathcal{O}(n)$, for X. Clearly $\delta\beta = 0$, so β is a cocycle. The 1-coboundaries, for this covering, are functions of the form $h_2 - h_1$, where h_2 is holomorphic on $\mathcal{U}_2$ and h_1, on $\mathcal{U}_1$. Calling $\mathcal{D} = \mathcal{U}_1 \cap \mathcal{U}_2$, $\mathcal{D}^- = \mathcal{U}_1$, $\mathcal{D}^+ = \mathcal{U}_2$, $h^- = -h_1$, and $h^+ = h_2$, we observe that the "equivalence" between twistor functions under the "gauge" freedom $\mathcal{G}$ that we started out with is just the normal cohomological equivalence between 1-cochains β, β' that their difference be a coboundary: $\beta' - \beta = \delta\alpha$, with $\alpha = (h_1, h_2)$. This suggests that we view the twistor function f as really defining us an element of $H^1(X, \mathcal{O}(n))$.

But this is all with respect to a particular covering of X, namely by $\{\mathcal{U}_1, \mathcal{U}_2\}$. Is this covering "fine" enough? There actually is a technical problem here. We cannot (normally) arrange that $\mathcal{U}_1$ and $\mathcal{U}_2$ are Stein manifolds (and in those exceptional cases when we can so arrange this, we would lose the invariance properties that we are striving for). It turns out, in fact, that this problem is not serious. One can show by direct construction (using the inverse twistor function--in the cases $n \leq -2$ at least, but probably in all cases) that for any <u>given</u> (analytic, positive frequency) field such a covering by two such sets $\mathcal{U}_1, \mathcal{U}_2$ is sufficient in the case $X = \overline{\mathbb{P}\mathbb{T}^+}$. Note that though X is invariant under SU(2,2), in this case, the <u>covering</u> is not. However, the cohomology group $H^1(X, \mathcal{O}(n))$ <u>is</u> invariant. Let us illustrate this by adding two elements of $H^1(X, \mathcal{O}(n))$ one of which is defined by a twistor function f, with respect to the covering $\{\mathcal{U}_1, \mathcal{U}_2\}$, and the other by $\hat{f}$, with respect to $\{\hat{\mathcal{U}}_1, \hat{\mathcal{U}}_2\}$, the second covering being a rotated version of the first. Schematically:

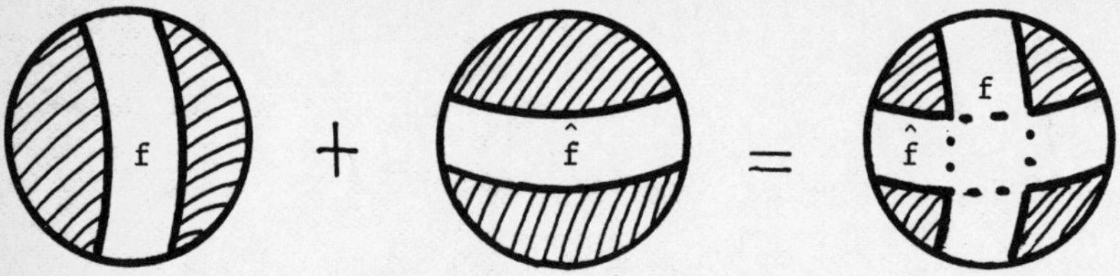

We define a representative cochain for the sum by taking the common refinement of both coverings. (Denote $\hat{\hat{\mathcal{U}}}_1 = \mathcal{U}_1 \cap \hat{\mathcal{U}}_1$, $\hat{\hat{\mathcal{U}}}_2 = \mathcal{U}_1 \cap \hat{\mathcal{U}}_2$, $\hat{\hat{\mathcal{U}}}_3 = \mathcal{U}_2 \cap \hat{\mathcal{U}}_1$, $\hat{\hat{\mathcal{U}}}_4 = \mathcal{U}_2 \cap \hat{\mathcal{U}}_2$ to give the refined covering $\{\hat{\hat{\mathcal{U}}}_1, \hat{\hat{\mathcal{U}}}_2, \hat{\hat{\mathcal{U}}}_3, \hat{\hat{\mathcal{U}}}_4\}$.) The 1-cocycle $\beta + \hat{\beta} = \hat{\hat{\beta}}$ is

$$(\hat{\hat{f}}_{12}, \hat{\hat{f}}_{13}, \hat{\hat{f}}_{23}, \hat{\hat{f}}_{14}, \hat{\hat{f}}_{24}, \hat{\hat{f}}_{34}) = (\hat{f}, f, f-\hat{f}, f+\hat{f}, f, \hat{f}).$$

Because of the "direct construction" argument mentioned above, this 1-cocycle will be cohomologous to (i.e. differing by a coboundary from) a cocycle of the form $(0, \hat{\hat{f}}, \hat{\hat{f}}, \hat{\hat{f}}, \hat{\hat{f}}, 0)$, so we can refer it back to the original covering if desired (although with $\mathcal{U}_1$ and $\mathcal{U}_2$ perhaps reduced slightly in size).

But we need not do so if we prefer not to. We have a generalization of the concept of a twistor function, namely as a collection of functions on portions of twistor space, defining a 1-cocycle with respect to some covering. We can actually use such a cocycle directly, obtaining the required space-time field by means of a branched contour integral. I shall just illustrate this with an example. Suppose that X is covered by 3 open sets $\mathcal{U}_1, \mathcal{U}_2, \mathcal{U}_3$, where on a projective line L in X we get the picture:

The 1-cocycle β is (f_{12}, f_{13}, f_{23}). To get the field $\phi\ldots.$ we perform the sum of three contour integrals with common end-points (in $\mathcal{U}_1 \cap \mathcal{U}_2 \cap \mathcal{U}_3$):

$$(2\pi i)^2 \phi\ldots. = \int_{\gamma_{12}} \ldots. f_{12} d^2 Z + \int_{\gamma_{13}} \ldots. f_{13} d^2 Z + \int_{\gamma_{23}} \ldots. f_{23} d^2 Z$$

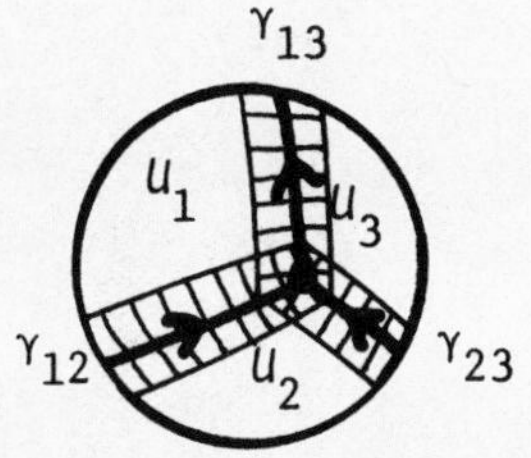

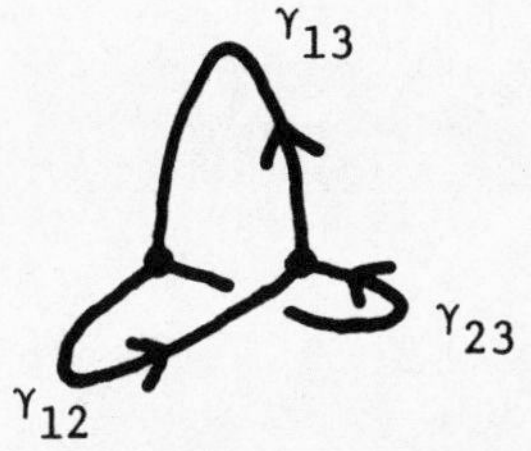

It is a simple matter to check that the cocycle condition $f_{12} - f_{13} + f_{23} = 0$ (in $\mathcal{U}_1 \cap \mathcal{U}_2 \cap \mathcal{U}_3$) ensures that the contour's end-points can be moved without affecting the result. This easily generalizes to coverings with N open sets.

Some applications (where knowledge of sheaf cohomology theory would be helpful).

(a) Charge integrality in the "twisted photon" description as opposed to vanishing charge in the description

$$\phi_{AB} = \frac{1}{(2\pi i)^2} \oint \frac{\partial}{\partial\omega^A} \frac{\partial}{\partial\omega^B} f(Z^\alpha) d^2\pi$$

(f homogeneous of degree 0).

We have [3] the exact sequence

$$0 \to \mathbb{Z} \xrightarrow{\times 2\pi i} \mathcal{O} \xrightarrow{\exp} \mathcal{O}^* \longrightarrow 0,$$

where $\mathcal{O}^*$ refers to non-zero holomorphic functions taken multiplicatively, from which we derive [3,4,5] the long exact sequence:

$$\begin{array}{ccccccccc} \ldots \longrightarrow & H^1(X,\mathbb{Z}) & \longrightarrow & H^1(X,\mathcal{O}) & \longrightarrow & H^1(X,\mathcal{O}^*) & \longrightarrow & H^2(X,\mathbb{Z}) & \longrightarrow \ldots \\ & \uparrow & & \uparrow & & \uparrow & & \uparrow & \\ & \text{ordinary} & & \phi_{..} = \oint \frac{\partial}{\partial\omega}\frac{\partial}{\partial\omega} f & & \text{"twisted} & & \text{ordinary} & \\ & \text{integer} & & & & \text{photon"} & & \text{integer} & \\ & \text{1st cohomology} & & & & & & \text{2nd cohomology} & \end{array}$$

Choose X as a region in $\mathbb{PT}$ corresponding to a small space-time tube surrounding a charge world-line.

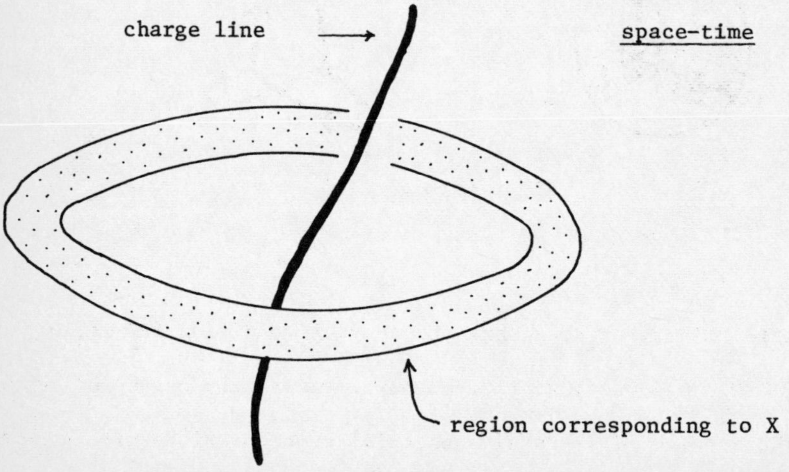

Then the topology of X is $S^2 \times S^2 \times R^2$, so $H^1(X,\mathbb{Z}) = 0$, and $H^2(X,\mathbb{Z}) \cong \mathbb{Z} \oplus \mathbb{Z}$. Thus the space $H^1(X,\mathcal{O}^*)$ effectively contains $H^1(X,\mathcal{O})$, and is strictly larger than $H^1(X,\mathcal{O})$ if the map to $\mathbb{Z} \oplus \mathbb{Z}$ is not simply to the zero element. The image in the first $\mathbb{Z}$ is always zero, but the image in the second $\mathbb{Z}$ is the value of the charge. (This much follows, for example, from examination of reference [6].) From this we see that the $H^1(X,\mathcal{O}^*)$ description only works if the charge has integer value (i.e. lies in $\mathbb{Z}$) whereas the $H^1(X,\mathcal{O})$ description only works if the charge value is zero. (In contrast to this, it may be remarked that there is no restriction on the charge value if the -4-homogeneity $f(Z^\alpha)$ description

$$\tilde{\phi}_{A'B'} = \frac{1}{(2\pi i)^2} \oint \pi_{A'}\pi_{B'} f(Z^\alpha) d^2\pi$$

is used, this corresponding to $H^1(X,\mathcal{O}(-4))$.)

(b) How to grow (or at least half-grow) new twistors from old.

There is another way of representing elements of $H^p(X,\mathcal{O})$, namely by taking the additive group of all $\bar{\partial}$-closed $(0,p)$-forms modulo the additive group of all $\bar{\partial}$-exact $(0,p)$-forms [3,5]. By a (q,p)-form is meant a $(q+p)$-form on X (not generally holomorphic) having q holomorphic differentials and p anti-holomorphic differentials:

$$\alpha_{i_1 \ldots i_q \, j_1 \ldots j_p} \, dz^{i_1} \wedge \ldots \wedge dz^{i_q} \wedge d\bar{z}^{j_1} \wedge \ldots \wedge d\bar{z}^{j_p} = \alpha .$$

Then $\bar{\partial}\alpha = \dfrac{\partial \alpha \ldots}{\partial \bar{z}^k} \, d\bar{z}^k \wedge dz^{i_1} \wedge \ldots \wedge d\bar{z}^{j_p}$ is a $\bar{\partial}$-exact $(q,p+1)$-form whose vanishing is the condition for α to be $\bar{\partial}$-closed. (We have $\bar{\partial}^2 = 0$.) Note that the $\bar{\partial}$-closed $(0,0)$-forms are holomorphic functions. We can also "twist" such forms or functions, as before.

Applying this to $H^1(X,\mathcal{O}(n))$ we obtain the result that the helicity $\frac{1}{2}(-n-2)$ massless fields can be described in terms of forms $g^{\alpha}(Z^{\cdot},\bar{Z}_{\cdot})d\bar{Z}_{\alpha}$ on X, modulo forms $\frac{\partial h(Z^{\cdot},\bar{Z}_{\cdot})}{\partial \bar{Z}_{\alpha}} d\bar{Z}_{\alpha}$, where $\frac{\partial g^{\alpha}}{\partial \bar{Z}_{\beta}} = \frac{\partial g^{\beta}}{\partial \bar{Z}_{\alpha}}$, the functions g^{α} being homogeneous of degree n in Z^{α} and of degree -1 in $\bar{Z}_{\alpha}$ (to balance the $d\bar{Z}_{\alpha}$). In order that the form $g^{\alpha}d\bar{Z}_{\alpha}$ be actually defined on the projective space (appropriately twisted) we require also $g^{\alpha}\bar{Z}_{\alpha} = 0$. The massless field integral now becomes

$$\phi_{A\ldots D} = \frac{1}{(2\pi i)^2}\oint \frac{\partial}{\partial\omega^{A}} \cdots \frac{\partial}{\partial\omega^{D}} g^{\gamma}(Z^{\cdot},\bar{Z}_{\cdot})d\bar{Z}_{\gamma} \wedge (I_{\alpha\beta}Z^{\alpha}dZ^{\beta})$$

or

$$\phi_{A'\ldots D'} = \frac{1}{(2\pi i)^2}\oint \pi_{A'}\cdots\pi_{D'}\, g^{\gamma}(Z^{\cdot},\bar{Z}_{\cdot})d\bar{Z}_{\gamma} \wedge (I_{\alpha\beta}Z^{\alpha}dZ^{\beta}).$$

(See also N. M. J. Woodhouse's article in Twistor Newsletter 2 for further details [7].)

There is presumably much freedom in the choice of g^{γ}. I shall suppose that its singularities can be arranged so that the contour in the above integrals (which is initially S^2, i.e., over the entire projective line corresponding to the space-time point in question) may be deformed, freeing $\bar{Z}_{\alpha}$ from Z^{α} in the process, until W_{α} ($=\tilde{Z}_{\alpha}$ = "freed" $\bar{Z}_{\alpha}$) passes through the line I. Writing $W_{\alpha} = I_{\alpha\beta}X^{\beta}$, we get

$$g^{\gamma}d\bar{Z}_{\gamma} \wedge I_{\alpha\beta}Z^{\alpha}dZ^{\beta} = g^{\rho}I_{\rho\sigma}dX^{\sigma} \wedge Z^{\alpha}I_{\alpha\beta}dZ^{\beta}.$$

The function g^{α} can be taken to be holomorphic in Z^{α} and W_{α}, with $W_{\alpha}g^{\alpha} = 0$ and $\frac{\partial g^{\alpha}}{\partial W_{\beta}} - \frac{\partial g^{\beta}}{\partial W_{\alpha}} = 0$. Restricting to $W_{\alpha} = I_{\alpha\beta}X^{\beta}$ we get $I_{\alpha\beta}g^{\alpha}X^{\beta} = 0$ whence

$g^{\alpha}I_{\alpha\beta} = qX^{\alpha}I_{\alpha\beta}$ for some $q(Z^{\alpha}, I_{\alpha\beta}X^{\beta})$ homogeneous of degrees (n,-2). Thus, the massless field integral can be written

$$\phi_{\dots} = \frac{1}{(2\pi i)^2} \oint_{S^2} \left\{ \begin{matrix} \frac{\partial}{\partial\omega}\dots \\ \text{or} \\ \pi\dots \end{matrix} \right\} qX^{\alpha}I_{\alpha\beta}dX^{\beta} \wedge Z^{\rho}I_{\rho\sigma}dZ^{\sigma},$$

which is the standard form for a 2-twistor integral. Thus we have conjured a new twistor out of thin air! This new twistor X^{α} is still only half-grown, though, since it appears as $I_{\alpha\beta}X^{\beta}$. This can be stated as $I^{\alpha\beta}\frac{\partial q}{\partial X^{\beta}} = 0$ which implies the massless relation $I_{\rho\sigma}Z^{\rho}X^{\sigma}I^{\alpha\beta}\frac{\partial^2 q}{\partial Z^{\alpha}\partial X^{\beta}} = 0$. In fact, the $\bar{\partial}$-closed property $\frac{\partial g^{\alpha}}{\partial W_{\beta}} - \frac{\partial g^{\beta}}{\partial W_{\alpha}} = 0$ now disappears on $W_{\alpha} = I_{\alpha\beta}X^{\beta}$, $g^{\alpha} \mapsto g^{\beta}I_{\beta\alpha}$, as it is incorporated in $I^{\alpha\beta}\frac{\partial q}{\partial X^{\beta}} = 0$ and $X^{\alpha}\frac{\partial q}{\partial X^{\alpha}} = -2q$ (homogeneity relation).

Acknowledgment

I am grateful to George Sparling, Andrew Hodges, and Nick Woodhouse for many helpful discussions and, most particularly, to Michael Atiyah for explaining sheaf cohomology theory so clearly to us and for many insightful remarks.

2. THE BACK-HANDED PHOTON (or, to a cricketer, THE GOOGLY PHOTON) *

Current twistor dogma presents the following picture of things:

(i) The wave function of a particle or system of particles is to be given by a "twistor function" $\psi(X^{\alpha},\dots,Z^{\alpha})$, which is holomorphic in all arguments $X^{\alpha},\dots,Z^{\alpha}$ --possibly with the modification that some (or all) of these should be dual twistors $W_{\alpha},\dots$ instead (or, conceivably, Hughston-type multi-twistors $X^{\alpha\beta\gamma},\dots$, etc.).

*Taken from Twistor Newsletter 3 (December, 1976).

(ii) More correctly, ψ is (presumably) an element of a sheaf cohomology group, say $\psi \in H^k(Q,\mathcal{O})$, for some suitable open set (or closure of an open set) $Q \subset \mathbb{T} \times \mathbb{T} \times \ldots \times \mathbb{T}$ say ($\mathbb{T}$ = twistor space), $\mathcal{O}$ being the sheaf of (germs of) holomorphic functions on Q (see previous section).

(iii) When a particle interacts it does so by deforming the structure of twistor space, or of $Q \subset \mathbb{T} \times \ldots \times \mathbb{T}$, whereby ψ now plays an active role in determining the precise deformation in question, thence affecting the behavior of other twistor functions defined on Q.

The above is vague in various respects. But something like (iii) is strongly suggested by two examples: the non-linear graviton [8] and the twisted photon [9]. (In fact, for gravitons, (iii) is strictly true only in the weak-field limit--unless some kind of non-linear sheaf theory can be evoked.) In the standard procedures for performing deformations (infinitesimal ones at least) it is only H^1's that are involved. Sparling has suggested that new procedures for performing deformations may be needed in order to be able to use H^k's ($k > 1$) in an active way. For the moment, I prefer the more modest idea that single particles are to be described by H^1's, these being the things that interact directly with other particles. H^2's, H^3's, etc. would play roles in describing many-particle states, with elements of suitable $H^k(Q,\mathcal{O})$ groups defining k-particle states. This could supply a possible answer to a question posed to me by Feynman: how do you know whether $\psi(X^\alpha,Y^\alpha,Z^\alpha)$ describes a hadron, or three massless particles, or a lepton and a massless particle? Elements of $H^1(Q,\mathcal{O})$, $H^3(Q,\mathcal{O})$, and $H^2(Q,\mathcal{O})$ are really quite different kinds of animals, after all! But the detailed implementation of this idea has proved elusive--not the least problem being to understand "twistorially" why the spin-statistics relation should hold for many-particle states.

Perhaps the most primitive problem, in trying to push forward with (iii), lies in the fact that a twisted photon is only <u>half</u> a photon, namely the left-handed half. Thus, the Ward(-Sparling) construction gives a deformation of $Q = \mathbb{T}^+$ starting from a twistor function $f(Z^\alpha)$, homogeneous of degree zero ($f \in H^1(\mathbb{T}^+, \mathcal{O})$ or $f \in H^1(\mathbb{P}\mathbb{T}^+, \mathcal{O}(0))$), to give a <u>left</u>-handed photon. Of course a right-handed photon can be produced by using $\tilde{f}(W_\alpha)$, of degree zero. But such would be to defeat the purpose of the economy of the twistor description (i). Simply changing the helicity quantum number (or any other quantum number) should not involve us in changing the <u>space</u> Q over which the twistor function is defined. Thus maintaining the general programme (i) - (iii) seems to lead us to the view that some form of deformation of $\mathbb{T}^+$ <u>must</u> be possible, which effectively encodes the information provided by an element $f \in H^1(\mathbb{T}^+, \mathcal{O})$, when f is homogeneous of degree -4 (i.e., in effect, $f \in H^1(\mathbb{P}\mathbb{T}^+, \mathcal{O}(-4))$. My suggestion for this, which relates closely to an earlier proposal due to Sparling and Ward [10], is as follows.

Consider, first, the standard Ward twisted photon. This is obtained by deforming the bundle (see figure) where the base space remains the undeformed

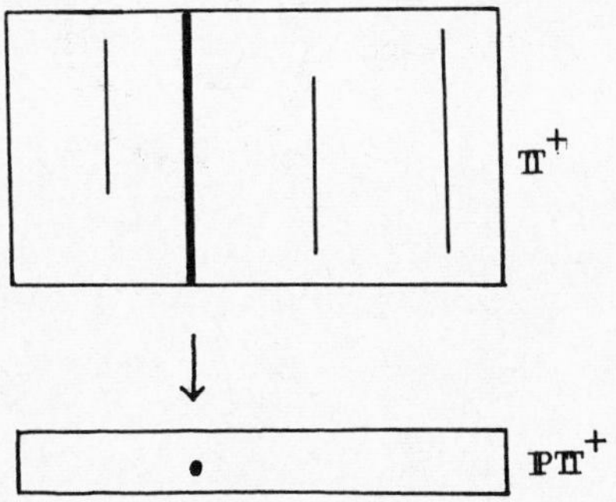

$\mathbb{P}\mathbb{T}^+$ and the fibre $\mathbb{C} - \{0\}$. Furthermore, the Euler operator $Z^\alpha \frac{\partial}{\partial Z^\alpha} =: T_Z$ remains globally defined. However, with non-trivial twisting, the forms $\mathbb{D}Z := \frac{1}{24} \varepsilon_{\alpha\beta\gamma\delta}\, dZ^\alpha \wedge dZ^\beta \wedge dZ^\gamma \wedge dZ^\delta$ and $\mathcal{D}Z := \frac{1}{6} \varepsilon_{\alpha\beta\gamma\delta}\, Z^\alpha dZ^\beta \wedge dZ^\gamma \wedge dZ^\delta$ are not well defined; indeed, globally defined analogues of these forms do not exist at all on a non-trivially twisted Ward photon. There is a good reason for this. In the case of flat $\mathbb{T}$, there is a canonical way of representing a twistor Z^α (up to the fourfold ambiguity $\pm Z^\alpha$, $\pm iZ^\alpha$) in terms of $\mathbb{P}\mathbb{T}$, namely as the projective twistor $\mathbb{Z}$ in $\mathbb{P}\mathbb{T}$, together with a 3-form at $\mathbb{Z}$ in $\mathbb{P}\mathbb{T}$. This 3-form lifts into $\mathbb{T}$ along the fibre over $\mathbb{Z}$; where it agrees with $\mathcal{D}Z$ defines us the point Z^α (or iZ^α, or $-Z^\alpha$, or $-iZ^\alpha$). Thus, if $\mathcal{D}Z$ is canonically known in the bundle, then the complete bundle structure is determined uniquely in terms of $\mathbb{P}\mathbb{T}$; therefore no twisting in the "photon" can occur. (This construction is the analogue of how one defines a spinor κ^A (up to sign) in terms of the celestial sphere.)

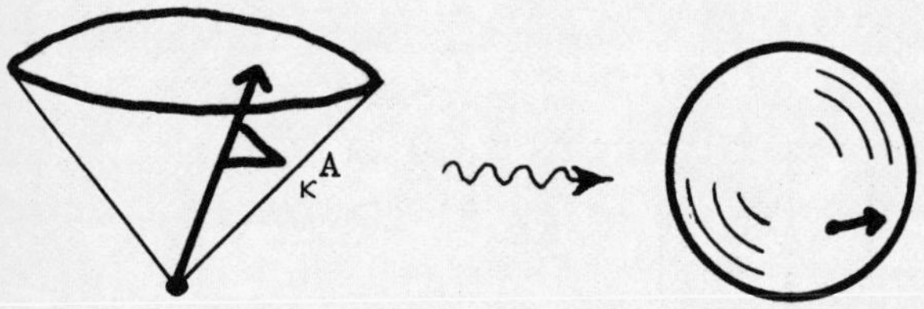

Furthermore, if a $\mathcal{D}Z$ exists globally in a bundle over $\mathbb{P}\mathbb{T}^+$, then it must be unique up to an overall constant factor.

Suppose, generally, we have a bundle (or holomorphic fibration) just <u>locally</u>, which is just a complex 4-space over a complex 3-space with fibre a complex 1-space. Then all we know locally in the 4-space is a direction

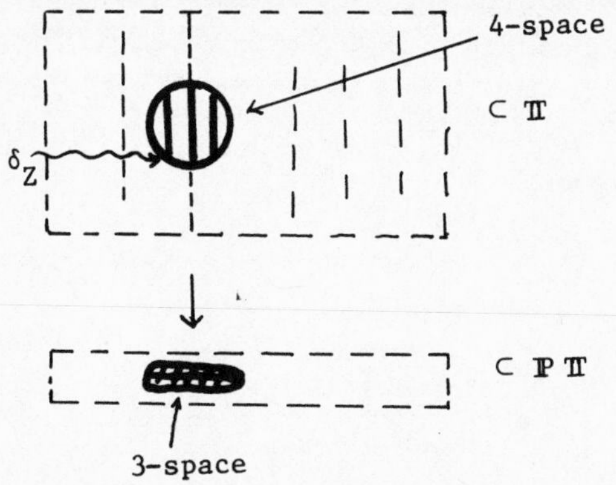

field δ_Z (1-foliation). To know $\mathcal{D}Z$ in the 4-space would be to know rather more structure than δ_Z. Being a 3-form, $\mathcal{D}Z$ is orthogonal to (i.e., annihilates) precisely one complex direction and so serves to <u>define</u> δ_Z. (I shall always assume $\mathcal{D}Z$ to be restricted to be orthogonal to δ_Z, so as to qualify to be a "$\mathcal{D}Z$.") Now $\mathcal{D}Z$ also defines a volume 4-form $\mathbb{D}Z$ by $d\mathcal{D}Z = 4\mathbb{D}Z$. Furthermore $\mathcal{D}Z$ and $\mathbb{D}Z$ together define the Euler operator T_Z, roughly speaking by "$T_Z = \mathcal{D}Z \div \mathbb{D}Z$," or more precisely by $\mathcal{D}Z = T_Z \lrcorner \mathbb{D}Z$. Conversely, this relation shows us that $\mathcal{D}Z$ is determined by the <u>pair</u> $(T_Z, \mathbb{D}Z)$. To know one or the other of T_Z, $\mathbb{D}Z$ is, by itself, not sufficient to determine $\mathcal{D}Z$, but the two together are equivalent to $\mathcal{D}Z$. Furthermore, assuming that T_Z is to point along δ_Z (where δ_Z is given), there is precisely as much information in T_Z locally as there is in $\mathbb{D}Z$. Each provides us with a kind of local scaling, but it is a "homogeneity degree 4" scaling for $\mathbb{D}Z$

and a "homogeneity degree 0" scaling for T_Z. In a sense $\mathbb{D}Z$ and T_Z seem to be sorts of duals to one another.

We can regard the Ward photon [9] as arising when we retain only the T_Z scaling and throw out $\mathbb{D}Z$. Let us try to do the "dual" thing and retain $\mathbb{D}Z$ while throwing out T_Z. I shall proceed in a fairly explicit way, assuming that two "coordinate" patches are given, where a standard twistor description is given in each patch, with X^α on the left and Z^α on the right. I am

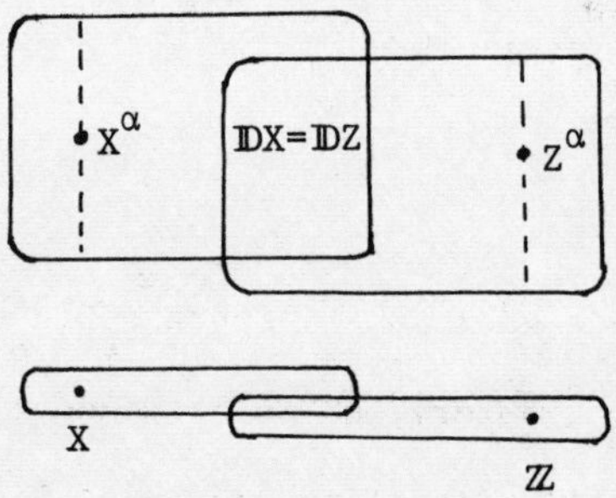

assuming there is no monkey-business in the base space, so $X^\alpha \propto Z^\alpha$ may be assumed on the overlap region. (The fibres in each half are given when X^α or Z^α is held constant up to proportionality.) The hypothesis is now $\mathbb{D}X = \mathbb{D}Z$ on the overlap (instead of Ward's $T_X = T_Z$), i.e., $d\mathcal{D}X = d\mathcal{D}Z$, i.e., $d\{\mathcal{D}X - \mathcal{D}Z\} = 0$. But $\delta_X = \delta_Z$, so $\mathcal{D}X \propto \mathcal{D}Z$. Put $\mathcal{D}X = \{1 + f(Z^\alpha)\}\mathcal{D}Z$; then $d\{f(Z^\alpha)\mathcal{D}Z\} = 0$, which holds iff $f(Z^\alpha)$ is <u>homogeneous of degree -4</u>. The transition relation is then

$$X^\alpha = \{1 + f_{-4}(Z^\alpha)\}^{\frac{1}{4}} Z^\alpha. \tag{A}$$

This appears to be a particular case of a Sparling-Ward [10] construction given earlier for massless fields of <u>any</u> helicity.

We run into trouble owing to branch points arising from the fourth root unless we exclude an extended region about the origin of each fibre. But things are okay near ∞ on the fibres. Thus, I envisage the fibres as being like $\mathbb{C}$ with some bounded (probably connected) region removed (i.e.,

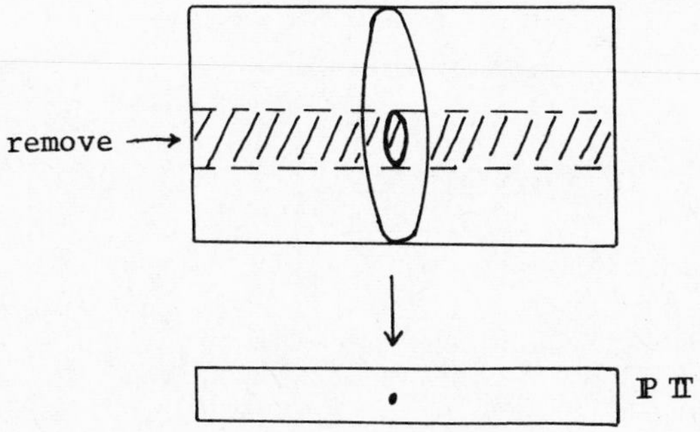

biholomorphic to $\mathbb{C} - \{z : |z| < 1\}$). Note that T_Z is not preserved by the patching, but related by $T_X = (1 + f(Z^\alpha))T_Z$ (which does strange things to the notion of a homogeneous function).

The intention is to regard $f(Z^\alpha)$ as a twistor function for a right-handed photon. But, in accordance with (ii), such $f(Z^\alpha)$ is really describing an element of a sheaf cohomology group. Thus we need to check that (A) is appropriately cohomological. Suppose we have a covering of the base space (say $\mathbb{P}\mathbb{T}^+$) by a number of open sets $\mathcal{U}_i$. Then on each $\mathcal{U}_i \cap \mathcal{U}_j$ we need $f_{ij} = -f_{ji}$ such that on each $\mathcal{U}_i \cap \mathcal{U}_j \cap \mathcal{U}_k$, $f_{ik} = f_{ij} + f_{jk}$ (cocycle condition). If we use standard twistor "coordinates" Z^α for the entire base space, then the f_{ij} will simply be homogeneous functions of Z^α, of degree -4. Thus, $f_{ij}(Z^\alpha) = -f_{ji}(Z^\alpha)$, $f_{ik}(Z^\alpha) = f_{ij}(Z^\alpha) + f_{jk}(Z^\alpha)$. The cocycle defining the right-handed photon can be taken to be just this collection of homogeneous functions. We piece together our "bundle" by taking "coordinates" Z_i^α for the portion lying above $\mathcal{U}_i$, where on each overlap (say, the portion of the bundle lying above $\mathcal{U}_i \cap \mathcal{U}_j$) we have $Z_i^\alpha = \{1 + f_{ij}(Z_j^\beta)\}^{\frac{1}{4}} Z_j^\alpha$. We must first check the consistency of this with $f_{ij} = -f_{ji}$. Avoiding messy indices we have (A) : $X^\alpha = \{1 + f(Z^\beta)\}^{\frac{1}{4}} Z^\alpha$ and $Z^\alpha = \{1 + g(X^\beta)\}^{\frac{1}{4}} X^\alpha$, and we need to check that $g = -f$. Now $Z^\alpha = \{1 + g(X^\beta)\}^{\frac{1}{4}}\{1 + f(Z^\beta)\}^{\frac{1}{4}} Z^\alpha$, so we require $(1 + g(X^\beta))(1 + f(Z^\beta)) = 1$; that is

$$1 = (1 + g(\{1 + f(Z^\beta)\}^{\frac{1}{4}} Z^\sigma))\,(1 + f(Z^\rho))$$

$$= (1 + \frac{g(Z^\sigma)}{1 + f(Z^\beta)})(1 + f(Z^\rho))$$

$$= (1 + f(Z^\rho) + g(Z^\rho))$$

(because of the -4 homogeneity of g) so $g(Z^\rho) = -f(Z^\rho)$ as required. Next we check the compatibility of (A) with $f_{ik} = f_{ij} + f_{jk}$. Again avoiding indices we have, on the triple intersection region above $\mathcal{U}_i \cap \mathcal{U}_j \cap \mathcal{U}_k$ relations

$$Y^\alpha = \{1 + p(X^\rho)\}^{\frac{1}{4}} X^\alpha; \quad X^\rho = \{1 + f(Z^\sigma)\}^{\frac{1}{4}} Z^\rho; \quad Y^\alpha = \{1 + q(Z^\sigma)\}^{\frac{1}{4}} Z^\alpha$$

For consistency we require

$$\{1 + p(\{1 + f(Z^\sigma)\}^{\frac{1}{4}} Z^\rho\}\{1 + f(Z^\beta)\} = 1 + q(Z^\sigma)$$

i.e. $$\{1 + \frac{p(Z^\rho)}{1 + f(Z^\rho)}\}\{1 + f(Z^\rho)\} = 1 + q(Z^\rho)$$

i.e. $q(Z^\rho) = p(Z^\rho) + f(Z^\rho)$ as required. We need also to show that if the cocycle is a coboundary, i.e., if $f_{ij} = f_j - f_i$ for each i,j, then the bundle is the same as that for flat twistor space (f_i being holomorphic throughout $\mathcal{U}_i$). For this we set $Z^\alpha = \{1 + f_i(Z_i^\beta)\}^{1/4} Z_i^\alpha$ and find that the Z^α, so constructed, is defined globally over the whole bundle. The compatibility over each $\mathcal{U}_i \cap \mathcal{U}_j$ follows by a calculation which is basically the same as the ones given above. Finally we need the fact that if the bundle is the same as that for flat twistor space (where here and above "the same" must be suitably interpreted in terms of "analytically extendible to"--owing to the gaping holes in the fibres), then the $\{f_{ij}\}$-cocycle is a coboundary. To prove this, we simply reverse the above argument. The flatness implies the existence of a global Z^α for the bundle. The local Z_i^α for each patch (above $\mathcal{U}_i$) must be related to Z^α by a formula $Z^\alpha = \{1 + f_i(Z_i^\beta)\}^{1/4} Z_i^\alpha$ from which follows (by the reverse of the above calculation) $f_{ij} = f_j - f_i$ above $\mathcal{U}_i \cap \mathcal{U}_j$.

The argument just given effectively shows that although (A) appears to be a highly non-linear relation, the system of bundles so constructed has nevertheless a linear structure (given by simply adding the sheaf cohomology group elements). This may be contrasted with Sparling's method of patching together non-linear gravitons:

$$X^\alpha = \exp\{I^{\rho\sigma} \frac{\partial f(Z^\tau)}{\partial Z^\rho} \frac{\partial}{\partial Z^\sigma}\} Z^\alpha.$$

For this, the cocycle condition fails. This is a manifestation of the very non-linearity of the non-linear graviton. On the other hand, the "4" in the above construction is not essential for linearity and could be replaced by other powers; i.e., f of homogeneity -n, with $X^\alpha = \{1 + f(Z^\beta)\}^{1/n} Z^\alpha$. When n=1, "0", -1, -2, -3,... this is implicit in the Sparling-Ward construction. However, when n < 0 the global structure of the fibres would have to be something different (because $(1 + z^{-n})^{1/n} z \to 1$ as $|z| \to \infty$ if n < 0, whereas

$(1 + z^{-n})^{1/n} z \sim z$ as $|z| \to \infty$ if $n > 0$). In any case, the motivation from "preservation of $\mathbb{D}Z$" exists only when $n = 4$.

A direct construction of the right-handed Maxwell field $\tilde{\phi}_{A'B'}$ (satisfying $\nabla^{AA'}\tilde{\phi}_{A'B'} = 0$) from the bundle structure has not yet emerged. But the most serious problem is that of fitting the right-handed and left-handed ways of deforming $\mathbb{T}^+$ together into one bundle...? One possibility for doing this is a method suggested by Hughston [12]. However it is unclear, as yet, how to extract left- and right-handed Maxwell fields which do not interact with one another. If this problem can be resolved, then among other things, it <u>might</u> suggest an analogous approach to an ambidextrous graviton!

3. REMARKS CONCERNING THE AMBIDEXTROUS PHOTON *

The information of a <u>left</u>-handed Maxwell field can be stored [9] in a holomorphic line bundle T over a suitable portion R of $\mathbb{P}\mathbb{T}$, the Euler operator T being well-defined on T and pointing along the fibres of T. Likewise, the information of a <u>right</u>-handed Maxwell field can be stored in a 1-manifold-fibration T over R, but where now it is the holomorphic volume 4-form $\Delta(=\mathbb{D}Z)$ which is well-defined on T instead of T. (See the previous section and also reference [9].) In reference [12], Hughston suggested putting the right-handed construction on top of the left-handed one, but it was not clear that this preserved any particular structure on T. In fact it does, namely the operator I shall call DIV. This is a map from vector fields V, pointing along the fibration, to scalars, satisfying $\mathrm{DIV}(\lambda V) = \lambda \mathrm{DIV}(V) + V(\lambda)$, where λ is a scalar field. It differs from an ordinary div operation only in that it need not act on any vector field which does not point along the fibration. Clearly, if Δ exists, then DIV exists, defined by $\mathrm{DIV}(V) = \mathrm{div}_\Delta V$, where

*Taken from Twistor Newsletter 5 (July, 1977).

div_Δ is the ordinary divergence based on Δ (i.e., use Δ to convert V into a 3-form $V \lrcorner \Delta$, take the exterior derivative of $V \lrcorner \Delta$, and then divide out by Δ to obtain a scalar). Also, if T exists, then DIV exists, defined by $T \times DIV(V) = [T,V] + 4V$.

Let us now suppose that the 4-dimensional space T, which is a fibration over R (all holomorphic), possesses just the structure DIV. Cover R by open sets $\{\mathcal{U}_i\}$ and lift these to give a covering $\{\mathcal{U}_i\}$ of T. Define a Δ_i and an T_i in each $\mathcal{U}_i$ which are compatible, in the sense above, with DIV. These compatibility requirements can be expressed as

$$DIV(T_i) = 4 \quad \text{and} \quad \mathcal{L}_{T_i}\Delta_i = 4\Delta_i$$

(the latter being equivalent to $div_{\Delta_i} T_i = 4$). Now define $f_{ij} = \log(\Delta_j/\Delta_i)$. We have $\mathcal{L}_{T_i} f_{ij} = 0$ and $f_{ij} - f_{ik} + f_{jk} = 0$ on $\mathcal{U}_i \cap \mathcal{U}_j \cap \mathcal{U}_k$, so f_{ij} defines a 1-cocycle on R. This is the left-handed (homogeneous of degree 0) "part" of the photon. Whenever this cohomology class vanishes, Δ exists globally. Suppose, next, that Δ _is_ global. Define $F_{ij} = (T_j - T_i) \lrcorner \Delta$. We find $\mathcal{L}_{T_i} F_{ij} = 0$ and $F_{ij} - F_{ik} + F_{jk} = 0$ on $\mathcal{U}_i \cap \mathcal{U}_j \cap \mathcal{U}_k$, so F_{ij} defines a 1-cocycle 3-form on R. This is the right-handed (back-handed)(homogeneous of degree -4) "part" of the photon.

In fact, this is exactly what we had before (in Twistor Newsletter 3 and [10]) although phrased somewhat differently, so that we don't even notice the $(\ldots)^{\frac{1}{4}}$ unpleasantness! (Incidentally, another way of disguising this is to write the $\hat{Z} = (1 + 4g)^{\frac{1}{4}} Z$ as $\hat{Z} = e^{gT}Z$, where $g(Z)$ is homogeneous of degree -4. The power series _do_ check!)

The trouble, of course, is the same as before: if there _is_ a left-handed part (Δ _not_ global) then there seems to be no way of extracting an

$H^1(R,\mathcal{O}(-4))$ element to measure the non-globalness of T (without bringing in more structure, that is). This is the old problem--the right-handed part appears to be charged with respect to the left-handed part. To decouple the two parts, it seems that strict conformal invariance must be broken (e.g. $F_{ij} = (T_j - T_i) \lrcorner \Delta_i$ or $F_{ij} = (T_j - T_i) \lrcorner (\Delta_j - \Delta_i)$ won't do because the cocycle condition fails). Thus we must bring in I (the "infinity twistor"). This works in principle, but is, as yet, inelegant.

4. MASSLESS FIELDS AND SHEAF COHOMOLOGY*

In the first of these four Twistor Newsletter articles I indicated that twistor functions are really to be regarded as providing representative cocycles for sheaf cohomology, so the twistorial representation of a massless field of "helicity $\frac{n}{2}$" is an element of $H^1(X,\mathcal{O}(-n-2))$, where X is an (open) region in $\mathbb{P}\mathbb{T}$ (say $\mathbb{P}\mathbb{T}^+$) swept out by the projective lines which correspond to the points of a suitable given (open) region Y in $\mathbb{C}\mathbb{M}$ (say $\mathbb{C}\mathbb{M}^+$) where the field is defined. Woodhouse [7] showed how the space-time field could be obtained from the $H^1(\ldots)$ element, when the $H^1(\ldots)$ element is defined by means of Dolbeault [3,4] cohomology (i.e., "anti-holomorphic" forms). A method due to Ward shows how to obtain the field from a Čech cohomology cocycle f_{ij} using an extension of Sparling's "splitting" formula (cf. Ward [9]). Ward's method essentially replaces the "branched contour integral" described in the first section (cf. also Ward [10]). Basically, Ward's method is as follows:

Let $f_{ij}(Z^\alpha)$ be homogeneous of degree -n-2 (with $n \geq 0$) and defined on $\mathcal{U}_i \cap \mathcal{U}_j$ (with $\{\mathcal{U}_i\}$ a covering of $X \subset \mathbb{P}\mathbb{T}$), $Z^\alpha = (\omega^A, \pi_{A'})$, $\omega^A = ix^{AA'}\pi_{A'}$, etc. Then $\pi_{A'} \ldots \pi_{E'} f_{ij}(Z^\alpha)$ (where π appears n+1 times) is homogeneous of

*Taken (with minor modifications) from Twistor Newsletter 5 (July, 1977).

degree -1 in π, so ("splitting") we get (ρ_i meaning "restriction," adopting a notational device due to Hughston, where square brackets denote skew-symmetrization):

$$\pi_{A'}\ldots\pi_{E'}f_{ij}(Z^{\alpha}) = \rho_{[i}\psi_{j]A'\ldots E'}(x^i,\pi_{R'})$$

whence

$$\phi_{jA'\ldots D'}(x,\pi) := \pi^{E'}\psi_{jA'\ldots D'E'}$$

satisfies $\rho_{[i}\phi_{j]A'\ldots D'} = 0$. Thus, due to the sheaf property

$\rho_{[i}F_{j]} = 0 \Rightarrow F_j = \rho_j G$, we have

$$\phi_{jA'\ldots D'}(x,\pi) = \rho_j\phi_{A'\ldots D'}$$

and, since $\phi_{A'\ldots D'}$ is homogeneous of degree 0 in π (and global in π), it must be a function of x only. Furthermore, $\nabla^{AA'}\phi_{A'B'\ldots D'} = 0$ readily follows. The case $n < 0$ is somewhat similar.

All this goes one way only: from twistor function to space-time field. But using the method of exact homology sequences we can readily extend this to obtain (implicitly) a version of the "inverse twistor function," which goes from space-time field to twistor function. (Compare also the Bramson-Sparling-Penrose method outlined in [1].) The following was inspired to some extent also by Hughston [17], cf. also Lerner [18].

Let $\mathbb{F}$ be the (dual) primed spin-vector bundle (excluding the section of zero spin vectors) over $\mathbb{CM}$. Thus, a point of $\mathbb{F}$ can be labelled $(x^a,\pi_{A'})$ (with $\pi_{A'} \neq 0$)--except for points at infinity on $\mathbb{CM}$--with x^a complex. $\mathbb{F}$ is also a bundle over $\mathbb{T} - \{0\}$, the fibre over Z^{α} being the set of linear 2-spaces containing Z^{α}.

$$\begin{array}{ccc} & \mathbb{F} & \\ \swarrow & & \searrow \\ \mathbb{T} - \{0\} & & \mathbb{CM} \end{array}$$

Consider the following exact sequence of sheaves over $\mathbb{F}$:

$$0 \longrightarrow T \xrightarrow{\times \pi_{A'}\cdots\pi_{E'}} Z'_{n+1} \xrightarrow{\times \pi^{E'}} Z'_n \longrightarrow 0. \tag{B}$$

Here Z'_n stands for the sheaf of n-index symmetric spinor fields $\phi_{A'\ldots D'}(x^i,\pi_{Q'})$ which are holomorphic in x^i and $\pi_{Q'}$ and satisfy the massless free-field equations $\nabla^{AA'}\phi_{A'B'\ldots D'} = 0$ ($\nabla_{AA'} = \frac{\partial}{\partial x^{AA'}}$, $\pi_{A'}$ being merely a passenger). The map $Z'_{n+1} \to Z'_n$ is simply

$\psi_{A'\ldots D'E'} \longmapsto \phi_{A'\ldots D'} = \pi^{E'}\psi_{A'\ldots D'E'}$ and we must check that is is <u>onto</u>, i.e., that given $\phi\ldots$ we can always solve for $\psi\ldots$. One method is as follows: <u>First</u>, fix $\pi_{A'}$, and choose a primed spinor basis $o^{A'}, \iota^{A'}$ with $\pi^{A'} = o^{A'}$. We have $\psi_{0'\ldots0'0'} = \phi_{0'\ldots0'}$, $\psi_{1'0'\ldots0'0'} = \phi_{1'0'\ldots0'}$, ..., $\psi_{1'\ldots1'0'} = \phi_{1'\ldots1'}$, but $\psi_{1'\ldots1'1'}$ is fixed only by the massless field equations

$$-\frac{\partial}{\partial x_{01'}}\psi_{1'\ldots1'1'} = \frac{\partial}{\partial x_{00'}}\psi_{1'\ldots1'0'} \quad \left(= \frac{\partial}{\partial x_{00'}}\phi_{1'\ldots1'}\right)$$

and

$$-\frac{\partial}{\partial x_{11'}}\psi_{1'\ldots1'1'} = \frac{\partial}{\partial x_{10'}}\psi_{1'\ldots1'0'} \quad \left(= \frac{\partial}{\partial x_{10'}}\phi_{1'\ldots1'}\right),$$

the integrability conditions for which come simply from the massless equations on $\phi_{A'\ldots D'}$. We can ensure a unique solution locally in $\mathbb{CM}$ by choosing (say) $\psi_{1'\ldots1'1'} = 0$ on an initial 2-surface $x_{11'} =$ constant, $x_{01'} =$ constant. <u>Second</u>, allow $\pi^{A'}$ to vary locally in $\mathbb{C}^2 - \{0\}$; take $\iota^{A'}$ to depend holomorphically on $\pi^{A'}$ ($= o^{A'}$) (say $\iota^{A'} \propto$ constant). The above prescription for finding $\psi_{A'\ldots D'E'}$ ensures that it is holomorphic in $\pi_{A'}$-- though the dependence on $\pi_{A'}$ may be complicated owing to the fact that the functional dependence of $\phi\ldots$ on the components $x_{\underline{A}0'}$ is π-dependent. (Incidentally, for simplicity, choose the unprimed basis constant.)

To see what T is we examine the kernel of the above map, i.e., find ψ's for which $\pi^{E'}\psi_{A'\ldots D'E'} = 0$ ($\pi^{E'} \neq 0$). We have (by standard lemmas) $\psi_{A'\ldots E'} = \pi_{A'}\ldots\pi_{E'}f$, and the massless equations on $\psi\ldots$ become $\pi^{A'}\nabla_{AA'}f = 0$. Thus the dependence of f on x^a is only through the quantity $\omega^A = ix^{AA'}\pi_{A'}$ (i.e., f is constant on twistor planes in $\mathbb{C}M$), so $f = f(\omega^A,\pi_{A'}) = f(Z^\alpha)$ is a twistor function. We see that T is the sheaf of twistor-type functions on $\mathbb{F}$; equivalently, of holomorphic functions on $\mathbb{T} - \{0\}$.

The short exact sequence (B) gives rise to the long exact homology sequence:

$$\ldots \to H^0(Q,Z'_{n+1}) \to H^0(Q,Z'_n) \to H^1(Q,T) \to H^1(Q,Z'_{n+1}) \to \ldots \qquad \text{(C)}$$

Now, restrict Z'_n to be homogeneous of degree 0 in $\pi_{A'}$ (sheaf $Z'_n(0)$); correspondingly, Z'_{n+1} to be homogeneous of degree -1 in $\pi_{A'}$ (sheaf $Z_{n+1}(-1)$); so that T will be homogeneous of degree -n-2 in $\pi_{A'}$ (sheaf $T(-n-2)$), i.e., of degree -n-2 in Z^α. Choose $Q \subset \mathbb{F}$ to correspond to the region of $\mathbb{F}$ lying above some $Y \subset \mathbb{C}M$ (so, for each $(x^a,\pi_{A'}) \in Q$ we have $(x^a,\hat{\pi}_{A'}) \in Q$ whenever $\hat{\pi}_{\underline{A'}} \in \mathbb{C}^2 - \{0\}$). Then since H^0 means "global sections" and since Z'_{n+1} is homogeneous of degree -1 in $\pi_{A'}$ we have

$$H^0(Q,Z'_{n+1}(-1)) = 0.$$

To study $H^1(Q, Z'_{n+1})$ we can use, for example, a "resolution" of Z'_{n+1} (where dependence on $\pi_{A'}$ is irrelevant)

$$0 \to Z'_{n+1} \to F_{0,n+1} \to F_{1,n} \to F_{0,n-1} \to 0, \qquad \text{(D)}$$

this being exact, where $F_{p,q}$ is the sheaf of fields $\chi^{A\ldots C}_{D'\ldots F'}(x^i, \pi_{R'})$ with p symmetric unprimed indices and q symmetric primed ones (freely holomorphic), so $F_{p,q}$ is a coherent analytic sheaf, yielding $H^r(S, F_{p,q}) = 0$ whenever S is Stein and r > 0. The maps are:

$$\phi_{A'\ldots E'} \longmapsto \chi_{A'\ldots E'} = \phi_{A'\ldots E'},$$

$$\chi_{A'B'\ldots E'} \longmapsto \theta^A{}_{B'\ldots E'} = \nabla^{AA'}\chi_{A'B'\ldots E'},$$

$$\theta^A{}_{B'C'\ldots E'} \longmapsto \eta_{C'\ldots E'} = \nabla^{B'}_A \theta^A{}_{B'C'\ldots E'},$$

and exactness is not hard to verify. If n=0, the sequence terminates one step sooner (and if n=-1 we also have a short exact sequence

$$0 \to Z'_0 \xrightarrow{i} F_{0,0} \xrightarrow{\Box} F_{0,0} \to 0$$

where $\Box$ is the D'Alembertian). Note that (D) is conformally invariant, the conformal weights of $\phi\ldots$, $\chi\ldots$, $\theta\ldots$, $\eta\ldots$ being, respectively, -1, -1, -3, -4. The sequence (D) (for which $\pi_{A'}$ is irrelevant) is of interest in the study of massless fields quite independently of twistor theory. Here we just use it to derive the fact that

$$H^1(Q, Z'_{n+1}) = 0$$

whenever Y is Stein (as is the case if Y is the future- (or past-) tube in $\mathbb{C}M$, or if Y is a suitable "thickening" of a portion of the real space-time

M. (This follows because $H^1(Q,F_{0,n+1}(-1))$ then vanishes from the homogeneity degree -1 in $\pi_{A'}$, the fact that $H^1(\mathbb{CP}_1,\mathcal{O}(-1)) = 0$ and suitable general theorems--for which thanks are due to R. O. Wells--and because $H^0(Q,F_{0,n+1}(-1)/Z_n'(-1))$ also vanishes--more obviously, because of -1 degree homogeneity.)

The upshot of this is that since the first and last terms in (C) both vanish (Y Stein), we have (for $n \geq 0$)

$$H^0(Q,Z_n'(0)) \cong H^1(Q,T(-n-2))$$

which, because globality in π on the left-hand-side implies constancy in π, almost establishes the required isomorphism between massless fields and H^1's of twistor functions. The remaining essential problem (the non-triviality and solution of which have been pointed out to me by M. Eastwood) is that all the sheaves refer to the space $\mathbb{F}$, so far, and some subtleties are involved in projecting the cohomology groups down to $\mathbb{T}$. The details of this and other matters are left to a proposed later paper to be written jointly with M. Eastwood and R. O. Wells, Jr. In fact the argument can be shown to work not only for the future tube (i.e., for $X = \mathbb{P}\mathbb{T}^+$) but also for suitable open regions in M. (Another subtlety has been glossed over in that the (x,π) description doesn't work for points at infinity. But this is unimportant because of conformal invariance and the fact that infinity can be transformed to somewhere safely inside the singular region of the field.)

To deal with the cases $n < 0$ we need a different exact sequence:

$$0 \rightarrow P_{m-2} \xrightarrow{i} T \xrightarrow{\frac{\partial}{\partial\omega^A}\cdots\frac{\partial}{\partial\omega^C}} Z_{m-1} \xrightarrow{\pi^{D'}\nabla_{DD'}} Z_m \rightarrow 0. \qquad \text{(E)}$$

Here Z_m stands for the sheaf of unprimed massless holomorphic fields $\phi_{A\ldots D}(x^i,\pi_{R'})$ with m = -n indices, and P_{m-2} stands for twistor functions which are polynomials in ω^A of degree at most m-2 for each fixed $\pi_{A'}$. The fact that $Z_{m-1} \to Z_m$ is onto, is basically an argument given if reference [11]; the kernel is this: twistor functions $\psi_{A\ldots C}(Z^\alpha)$, the massless field equations on which are $\pi^{E'}\frac{\partial}{\partial\omega^{[D}}\psi_{A]\ldots C} = 0$, yielding $\psi_{A\ldots C} = \frac{\partial}{\partial\omega^A}\cdots\frac{\partial}{\partial\omega^C} f(Z)$ as required. The short exact sequence

$$0 \to T/P_{m-2} \to Z_{m-1} \to Z_m \to 0$$

provides the long exact sequence

$$H^0(Q,Z_{m-1}) \to H^0(Q,Z_m) \to H^1(Q,T/P_{m-2}) \to H^1(Q,Z_{m-1}). \qquad \text{(F)}$$

Choosing homogeneities -1, 0, m-2, -1, respectively, and using the same argument as above, we get (Y Stein):

$$H^0(Q,Z_m(0)) \cong H^1(Q,T(m-2)/P_{m-2}(m-2)). \qquad \text{(G)}$$

To deal with the extra complication of P_{m-2} we use $0 \to P_{m-2} \to T \to T/P_{m-2} \to 0$ to derive the exact sequence

$$H^1(Q,P_k(k)) \to H^1(Q,T(k)) \to H^1(Q,T(k)/P_k(k)) \to H^2(Q,P_k(k)) \qquad \text{(H)}$$

(where k = m-2). Now the sheaf $P_k(k)$ vanishes unless $k \geq 0$ (i.e., $m \geq 2$) and, using a suitable Kunneth formula, we derive the fact (since homogeneity in $\pi_{A'}$ for each x is non-negative) that we are concerned only with polynomial behavior in $\pi_{A'}$ for the two end terms in (H). Consequently these terms refer, in effect, to the "ordinary" cohomology of Q but with coefficients which are twistor polynomials $Z^\alpha\ldots Z^\gamma P_{\alpha\ldots\gamma}$, homogeneous of degree k, i.e., equivalently, symmetric n-twistors $P_{\alpha...\gamma}$. If Y describes a space-time region surrounding a "source tube," then we have $\frac{1}{6}(k+1)(k+2)(k+3) = \frac{1}{6}m(m^2-1)$ independent complex "charges," corresponding to the various independent components of $P_{\alpha...\gamma}$, all of which have to vanish if the $H^1(Q,\mathcal{T})$ description is to work. This corresponds to the final map in (H) mapping to the zero element. (Note that m=2 for the anti-self-dual Maxwell case, which is consistent with a statement made in the first section of this article.) If Y is the forward tube $\mathbb{CM}^+$, then the first and last terms of (H) both vanish and the required isomorphism follows (but various details remain to be worked out).

Alternative approaches to the cases $n < 0$ can be given in which a potential rather than the field is used. This avoids having to bring in $\mathcal{P}_k$ and $P_{\alpha...\gamma}$. (Work by Ward, Hughston and Eastwood.)

Further work is in progress.

Thanks are due to M. F. Atiyah, M. Eastwood, R. O. Wells, Jr. and R. S. Ward.

5. SOME REMARKS CONCERNING A GOOGLY* GRAVITON

The standard "non-linear graviton" construction [8] provides a means of coding a general anti-self-dual solution of Einstein's equations (or "right-flat space-time") into the structure of a deformed twistor space. A general self-dual solution (or "left-flat space-time") can be correspondingly coded into the structure of a deformed dual twistor space. However, it appears to be important, for the success of the twistor program as a whole, that the same twistor space be employable to encode the

*To those unfamiliar with the game of cricket, it should be explained that a "googly" is a ball bowled in such a way that it spins in a right-handed sense about its direction of motion even though, to the batsman, it would appear that the bowling action would be such as to impart a left-handed helicity.

information of *both* anti-self-dual and self-dual types of vacuum curvature. The hope would be that by combining both parts of the space-time (Weyl) curvature into the structure of *one* suitably deformed twistor space it might be possible to represent, twistorially, the general (analytic?) solution of Einstein's vacuum equations. I shall here be concerned mainly with the preliminary problem of attempting to encode the structure of a left-flat space-time into a deformed *twistor* space.

The essential difficulty in doing this appears to lie in the fact that the interpretation of a (complex) space-time point as a compact holomorphic curve in deformed projective twistor space, with null space-time separation being interpreted as intersection of the corresponding holomorphic curves, leads directly to the condition that many α-planes exist in the space-time and that consequently the space-time is right-conformally flat. What seems to be required is a different twistorial interpretation of a space-time point. Now recall [13] the simple exact sequence that expresses the essential Poincaré invariant structure of *flat* twistor space $\mathbb{T}$

$$0 \to S \xrightarrow{i} \mathbb{T} \xrightarrow{p} \tilde{\mathbb{S}}^* \to 0$$

where $\mathbb{S}$ is "unprimed" spin-space (the space of ω^A) and $\tilde{\mathbb{S}}^*$ is dual "primed" spin-space (the space of $\pi_{A'}$). The map i is $\omega^A \mapsto (\omega^A, 0)$ and p is $(\omega^A, \pi_{A'}) \mapsto \pi_{A'}$, with $Z^\alpha = (\omega^A, \pi_{A'})$. The standard twistorial representation of a (complex) space-time point may be thought of as a cross-section of $\mathbb{T}$, where $\mathbb{T}$ is regarded as a bundle over $\tilde{\mathbb{S}}^*$ with projection p, the cross-section being lined up along the direction of the Euler operator $T = Z^\alpha \partial/\partial Z^\alpha$ and hence passing through the origin of $\mathbb{T}$. Another way of viewing this cross-section is as a map q from $\tilde{\mathbb{S}}^*$ to $\mathbb{T}$ such that the composition $p \circ q$ is the identity on $\tilde{\mathbb{S}}^*$.

This is one method of "splitting" the exact sequence; there is a corresponding dual method which is simply a map j from $\mathbb{T}$ to $ with the property that the composition j∘i is the identity on $. In the flat case, the maps j and q contain the same information as each other, so they can each equivalently be used to represent a (complex) space-time point. However, when $\mathbb{T}$ is suitably (mutilated and) deformed, the information will be different. The map q now gives the standard representation of a point in the right-flat non-linear graviton construction; and some corresponding "deformed" version of the map j ought to give a different difinition of a "point," the space of these "points" providing the required left-flat complex space-time.

There are many difficulties in making such an idea work in detail, however. It seems that we need some non-compact version of Kodaira's theorem [14,15] in order to be sure that the family of deformed j-maps is actually 4-dimensional. We need some suitable (and yet unknown) regularity conditions in the neighborhood of the image of i. Indeed, it appears that this neighborhood contains the entire information of the self-dual curvature. We need, also, to be able to see why this neighborhood, if only infinitesimally deformed, is characterized by a 1st cohomology group of $\mathcal{O}(-6)$ twistor functions. Work is in progress on these problems.

The idea, in order to proceed further, would be that with one and the same deformed twistor space we might have three alternative definitions of "space-time point," the above two and a third which would be some kind of symmetrical combination of the other two, providing a general solution of Einstein's vacuum equations! Assuming this solution to be asymptotically flat, the q-definition would provide its $\tilde{H}$-space and the p-definition its H-space [16]. At the moment, however, this is all purely speculative.

References

1 R. Penrose, "Twistor Theory, Its Aims and Achievements" in Quantum Gravity, eds: C. J. Isham, R. Penrose, D. W. Sciama, Oxford University Press, 1975.

2 R. Penrose, "Twistors and Particles" in Quantum Theory and the Structures of Time and Space, eds: L. Castell, M. Drieschner, C. F. von Weiszacker, C. Hanser, Munich 1975.

3 J. Morrow and K. Kodaira, Complex Manifolds; Holt, Rinehart and Winston, N. Y., 1971.

4 R. C. Gunning and H. Rossi, Analytic Functions of Several Complex Variables, Prentice-Hall, 1965.

5 R. O. Wells, Jr., Differential Analysis on Complex Manifolds, Prentice-Hall, 1973.

6 G. A. J. Sparling and R. Penrose, "The Twistor Quadrille" in Twistor Newsletter 1(4 March 1976).

7 N. M. J. Woodhouse, "Twistor Cohomology Without Sheaves," in Twistor Newsletter 2(10 June 1976).

8 R. Penrose Gen. Rel. Grav. 7, 31, 171(1976).

9 R. S. Ward, "The Twisted Photon" in Twistor Newsletter 1(4 March 1976).

10 R. S. Ward, "Zero-rest-mass Fields from Twistor Functions" in Twistor Newsletter 1 (4 March 1976).

11 R. Penrose, Proc. Roy. Soc. A284, 159(1965).

12 L. P. Hughston, "A Generalized Right-Handed Photon Construction" in Twistor Newsletter 3 (7 December 1976).

13 R. Penrose and M. A. H. MacCallum, Phys. Reports 6C, 241(1973).

14 K. Kodaira, Am. J. Math. 85, 79(1963).

15 K. Kodaira, Ann. Math. 75, 146(1962).

16 R. Hansen, E. T. Newman, R. Penrose and K. P. Tod, Proc. Roy. Soc. (to appear).

17 L. P. Hughston, "The Twistor Cohomology of Local Hertz Potentials" in Twistor Newsletter 4 (1 April 1977).

18 D. Lerner, "The 'Inverse Twistor Function' for Positive Frequency Fields" in Twistor Newsletter 5(11 July 1977).

R PENROSE
Mathematical Institute
24-29 St. Giles
Oxford, England

R O Wells, Jr

Cohomology and the Penrose transform

Penrose has described in various places his remarkable correspondence between Minkowski space and subsets of $\mathbb{P}_3(\mathbb{C})$ first introduced in [5]. We will recall briefly how the correspondence can be described in terms of a basic double fibration [12]. Consider complex flag manifolds defined by complex subspaces of various dimensions of $\mathbb{C}^4$. Namely, let

$$\mathbb{P} = \{S_1 \subset \mathbb{C}^4 : \dim S_1 = 1\},$$
$$\mathbb{M} = \{S_2 \subset \mathbb{C}^4 : \dim S_2 = 2\},$$
$$\mathbb{F} = \{S_1 \subset S_2 \subset \mathbb{C}^4 : \dim S_1 = 1, \dim S_2 = 2\}.$$

Then $\mathbb{P}$, $\mathbb{M}$, and $\mathbb{F}$ are compact complex manifolds of dimensions 3, 4, and 5, respectively, with $\mathbb{P}$ being a projective space, $\mathbb{M}$ a Grassmannian manifold and $\mathbb{F}$ a more general flag manifold. There is a natural double fibration of holomorphic bundle mappings

$$\begin{array}{ccc} & \mathbb{F} & \\ {\scriptstyle\alpha}\swarrow & & \searrow{\scriptstyle\beta} \\ \mathbb{P} & \overset{\tau}{\rightsquigarrow} & \mathbb{M} \end{array} \qquad (1)$$

inducing the Penrose correspondence τ between $\mathbb{P}$ and $\mathbb{M}$ defined by

$$\tau(p) = \beta(\alpha^{-1}(p)) \cong \mathbb{P}_2(\mathbb{C}), \quad p \in \mathbb{P}$$
$$\tau^{-1}(p) = \alpha(\beta^{-1}(p)) \cong \mathbb{P}_1(\mathbb{C}), \quad p \in \mathbb{M}.$$

Thus points in $\mathbb{P}$ correspond to projective planes in $\mathbb{M}$ and points in $\mathbb{M}$ correspond to projective lines in $\mathbb{P}$.

We now introduce an Hermitian quadratic form Φ on $\mathbb{C}^4$ of signature (++--). The pair $(\mathbb{C}^4, \Phi)$ is called the space of *twistors*, and is denoted by $\mathbb{T}$. The compact complex manifolds above have homogeneous coordinates defined in terms of spinors and these coordinates are important for many calculations involving function-theoretic and field-theoretic quantities on these manifolds. The conceptual results we want to describe will not normally involve these coordinates, but certain details of our proofs and the physical interpretation of some of our results certainly involve the twistor coordinates a great deal. We now use the twistor metric Φ to define various subsets of our three complex manifolds $\mathbb{P}$, $\mathbb{F}$, and $\mathbb{M}$. We will say that a subspace $S \subset \mathbb{C}^4$ is: a) *positive* if $\Phi(v)$ is positive for any nonzero vector $v \in S$, b) negative if $\Phi(v)$ is *negative* for any nonzero vector $v \in S$, and c) *null* if $\Phi(v)$ is zero for any vector $v \in S$. We now define

$$\mathbb{P}_+ = \{S_1 \subset \mathbb{C}^4 : \dim S_1 = 1, S_1 \text{ positive}\},$$
$$\mathbb{P}_- = \{S_1 \subset \mathbb{C}^4 : \dim S_1 = 1, S_1 \text{ negative}\},$$
$$P = \{S_1 \subset \mathbb{C}^4 : \dim S_1 = 1, S_1 \text{ null}\},$$

with $\mathbb{M}_+$, $\mathbb{M}_-$, M, and $\mathbb{F}_+$, $\mathbb{F}_-$, and F defined in a similar manner. Thus we have induced diagrams:

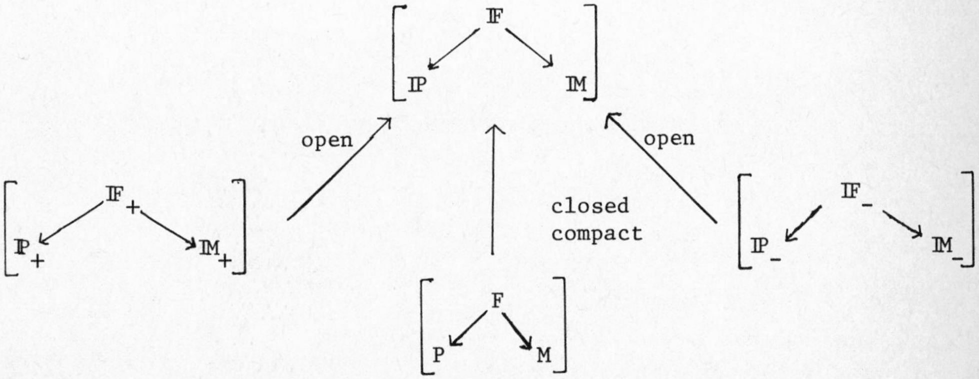

with the corresponding induced Penrose correspondences. We find that M is compactified real Minkowski space, and that SU(2,2), which acts transitively on the proper subsets $\mathbb{F}_+$, etc., defined above by Φ, induces on M the action of the conformal extension of the Poincare group. Moreover, M is a boundary component of both $\mathbb{M}_+$ and $\mathbb{M}_-$ which are bounded symmetric domains, and $\mathbb{M}$ is a complexification of M. Similarly, F is a 6-real-dimensional submanifold of $\mathbb{F}$ which is a CR-submanifold of CR-dimension 1, i.e. the holomorphic tangent space of F in $\mathbb{F}$ is 1-dimensional (cf. [11]), and moreover, F is a boundary component of both $\mathbb{F}_+$ and $\mathbb{F}_-$. In the same manner, P is a 5-real-dimensional hypersurface in $\mathbb{P}$, and has CR-dimension 2. For more details concerning these submanifolds and their construction see [12]. We remark that $\mathbb{F}$, $\mathbb{F}_+$, $\mathbb{F}_-$, $\mathbb{P}$,..., all are complex manifolds, while P, F, and M are CR-manifolds which means they have a certain amount of tangential complex structure inherited from their ambient complex manifolds $\mathbb{P}$, $\mathbb{F}$, and $\mathbb{M}$.

Our main problem is to describe solutions of various massless field equations on M in terms of intrinsic holomorphic data on $\mathbb{P}$ or subsets of $\mathbb{P}$. At the conclusion of [12], we indicated how holomorphic massless fields of positive helicity on $\mathbb{M}_+$ could be described in terms of cohomology classes on $\mathbb{P}_+$, a result due to R. Penrose (cf. [7]). In this paper we want to indicate how this description can be extended to negative helicity and to weak (hyperfunction) solutions on the real compactified Minkowski space M. We will first recall the construction of [12] for positive helicity which necessitates introducing some terminology from the theory of complex manifolds (cf. [10]).

Let X be a complex manifold and let $\mathcal{O}_X$ be the sheaf of (local) holomorphic functions on X, the <u>structure sheaf</u> of X. If $V \to X$ is a holomorphic vector bundle, then we will let $\mathcal{O}_X(V)$ be the sheaf of (local) holomorphic sections of V. One sees easily that $\mathcal{O}_X(V)$ is a locally free sheaf of modules over the sheaf of rings $\mathcal{O}_X$, and that any locally free sheaf of modules over $\mathcal{O}_X$ is the sheaf of holomorphic sections of a holomorphic vector bundle. Now consider a holomorphic mapping $X \xrightarrow{f} Y$ of two complex manifolds. If $\mathcal{F} = \mathcal{O}_Y(V)$ is a locally free sheaf on Y, then we define the pullback sheaf $f^*\mathcal{F}$ to be $\mathcal{O}_X(f^*V)$, the sheaf of holomorphic sections of the pullback bundle. On the other hand if $\mathcal{G}$ is any sheaf of abelian groups on X, then we define a sequence of sheaves $\{f_*^q\mathcal{G}\}$ on Y called the direct image sheaves under the mapping f. The qth direct image sheaf of $\mathcal{G}$ under f, denoted by $f_*^q\mathcal{G}$, is the sheaf generated by the presheaf

$$U \to H^q(f^{-1}(U), \mathcal{G}), \quad \text{for } U \text{ open in } Y.$$

The stalk of $f_*^q\mathcal{G}$ at $P \in Y$ is the direct limit

$$(f_*^q\mathcal{G})_p = \varinjlim_{U \ni p} H^q(f^{-1}(U), \mathcal{G}),$$

which is essentially the cohomology along the fiber $f^{-1}(p)$ (see [2], for a discussion of direct image sheaves). We want to use these two basic constructions of lifting and pushing down sheaves in conjunction with our double fibration (1) to construct certain fundamental sheaves on $\mathbb{P}$, $\mathbb{F}$, and $\mathbb{M}$.

Let $n \in \mathbb{Z}$, and let $H^{-n-2} \longrightarrow \mathbb{P}$ be the hyperplane section bundle of $\mathbb{P} = (\mathbb{P}_3(\mathbb{C}))$ raised to the power -n-2, i.e., the local sections of H^{-n-2} are homogeneous functions in the homogeneous coordinates of $\mathbb{P}$ of degree -n-2. We define, for $n \in \mathbb{Z}$,

$$\begin{aligned} \mathcal{S}_{\mathbb{P}}(n) &= \mathcal{O}_{\mathbb{P}}(H^{-n-2}), \\ \mathcal{S}_{\mathbb{F}}(n) &= \alpha^*\mathcal{O}_{\mathbb{P}}(H^{-n-2}), \qquad (2) \\ \mathcal{S}_{\mathbb{M}}(n) &= \beta_*^1\alpha^*\mathcal{O}_{\mathbb{P}}(H^{-n-2}). \end{aligned}$$

We note that we only use the first direct image sheaf in (2), since it turns out that (as described later in the paper)

$$\beta^j_* \alpha^* \mathcal{O}_{\mathbb{P}}(H^{-n-2}) = 0, \text{ for } j \neq 1.$$

Thus the sheaves $S_{\mathbb{F}}(n)$ and $S_{\mathbb{M}}(n)$ are the natural pullback and pushforward of the sheaf $S_{\mathbb{P}}(n)$ in this context. One can show that, if $n \geq 0$,

$$S_{\mathbb{M}}(n) = \mathcal{O}_{\mathbb{M}}(V_n),$$

where

$$\text{rank}_{\mathbb{C}} V_n = \text{rank} \odot^n (\mathbb{C}^2),$$

and where $\odot^n$ denotes the $n^{\underline{th}}$ symmetric tensor product. Moreover, sections of $S_{\mathbb{M}}(n)$ can be identified with holomorphic spinors of primed type on open subsets of $\mathbb{M}$, i.e., of the form $\{\phi_{\underbrace{A'B'\ldots D'}_{n}}\}$ (cf. e.g. [5] and [12]).

This is the basic local analytic data which will allow us to transform cohomology on subsets of $\mathbb{P}$ to spinor fields on subsets of $\mathbb{M}$. For $n < 0$, we will see that

$$\beta^1_* \alpha^* \mathcal{O}_{\mathbb{P}}(H^{-n-2}) = 0,$$

and our construction breaks down. We will consider the case $n < 0$ later on in this paper, and concentrate for the moment on the case $n > 0$ (which will correspond to positive helicity from the point of view of physics). Thus we have three basic sheaves on our three basic spaces, depending on the integer n. On $\mathbb{P}$ we have "homogeneous functions" $S_{\mathbb{P}}(n)$; on $\mathbb{F}$ we have "pullbacks of homogeneous functions" $S_{\mathbb{F}}(n)$, and on $\mathbb{M}$ we have "spinors" $S_{\mathbb{M}}(n)$, canonically related as sheaves by the geometry of the basic diagram (1).

We need one more ingredient, namely, the differential operators defined by the geometry of (1) which will correspond to certain field equations arising in classical mathematical physics. Let $T(\mathbb{F})$ be the tangent bundle to $\mathbb{F}$, and let $T_\alpha(\mathbb{F}) \subset T(\mathbb{F})$ be the subbundle of vectors tangent to the fibers of the fibration α. This induces a canonical surjection of the dual bundles

$$T^*(\mathbb{F}) \xrightarrow{\pi_\alpha} T_\alpha^*(\mathbb{F}),$$

and a differential operator

$$d_\alpha = \pi_\alpha \circ d$$

acting on differential forms of any degree. Thus if we have smooth differential forms of degree r on $\mathbb{F}$, denoted by $\mathcal{E}^r(\mathbb{F})$, then exterior differentiation can be considered as a mapping

$$\mathcal{E}^r(\mathbb{F}) \xrightarrow{d} \mathcal{E}^r(\mathbb{F}, T^*(\mathbb{F}))$$

of scalar r-forms to $T^*(\mathbb{F})$-valued r-forms. Thus we obtain by composition with π_α a differential operator

$$\mathcal{E}^r(\mathbb{F}) \xrightarrow{d_\alpha} \mathcal{E}^r(\mathbb{F}, T_\alpha^*(\mathbb{F})).$$

This differential operator corresponds to "differentiation along the fibers of α," and will play an important role below. We want to use $T_\alpha^* = T_\alpha^*(\mathbb{F})$ to introduce two new sheaves on $\mathbb{F}$ and $\mathbb{M}$ respectively. Namely, consider

$$\mathcal{S}_{\mathbb{F}}^{\alpha}(n) = \mathcal{O}_{\mathbb{F}}(\alpha^* H^{-n-2} \otimes T_\alpha^*),$$
$$\mathcal{S}_{\mathbb{M}}^{\alpha}(n) = \beta_*^1 \mathcal{O}_{\mathbb{F}}(\alpha^* H^{-n-2} \otimes T_\alpha^*),$$

i.e., the extension of our basic sheaves on $\mathbb{F}$ and $\mathbb{M}$ by the cotangent bundle along the fibers of α. This data allows us to construct the following diagram, where all of the sheaves depend on the integer n, assumed positive for now:

$$\begin{array}{ccccc} & \overset{\alpha^*}{\nearrow} & H^1(\mathbb{F}_+, \mathcal{S}_{\mathbb{F}}(n)) & \xrightarrow{d_\alpha} & H^1(\mathbb{F}_+, \mathcal{S}^\alpha_{\mathbb{F}}(n)) \\ H^1(\mathbb{P}_+, \mathcal{S}_{\mathbb{P}}(n)) & & \cong \downarrow I & & \cong \downarrow I_\alpha \\ & \underset{p}{\searrow} & H^0(\mathbb{M}_+, \mathcal{S}_{\mathbb{M}}(n)) & \xrightarrow{\nabla_\alpha} & H^0(\mathbb{M}_+, \mathcal{S}^\alpha_{\mathbb{M}}(n)) \end{array} \tag{3}$$

The differential operator d_α extends to bundle-valued differential forms on $\mathbb{F}$ since the bundle $\alpha^* H^{-n-2}$ has transition functions which are constant along the fibers of α. It extends to cohomology since d_α (anti-)commutes with $\bar{\partial}$, and we are representing the sheaf cohomology in terms of differential forms via the Dolbeault isomorphism. The vertical mappings I and I_α are the canonical isomorphisms given by the Leray spectral sequence (cf. [2]),

$$H^q(\mathbb{M}_+, \beta^j_* \mathcal{S}_{\mathbb{F}}) \Rightarrow H^r(\mathbb{F}_+, \mathcal{S}_{\mathbb{F}})$$

which degenerates since $\beta^j_* \mathcal{S}_{\mathbb{F}}(n) = 0$, for $j \neq 1$. They are discussed in more detail later in the paper. Thus the differential operator d_α induces a first-order differential operator ∇_α in (3) which turns out to be the zero-rest-mass operator acting on spinor fields (cf. [12] and [1]). The mapping $p = I \circ \alpha^*$ which transforms cohomology in $\mathbb{P}_+$ to spinor fields on $\mathbb{M}_+$ is called the Penrose transform. One can show that p is one-to-one and that its image in $H^0(\mathbb{M}_+, \mathcal{S}_{\mathbb{M}}(n))$ is the same as the kernel of ∇_α. One thus has the basic isomorphism for $n > 0$,

$$H^1(\mathbb{P}_+, F_{\mathbb{P}}(n)) \underset{\cong}{\xrightarrow{p}} \{\text{holomorphic solutions of helicity } s = n/2 \text{ of the zero-rest-mass field equations on } \mathbb{M}_+\}. \tag{4}$$

See [12] for a further discussion of this isomorphism. The detailed proofs will appear in [1].

The basic ingredients of the Penrose transform and the isomorphism (4) are contained in diagram (3). We started with the fibration (1), and the hyperplane section bundle H on $\mathbb{P}$. We considered the sheaf $\mathcal{S}_{\mathbb{P}}(n) = \mathcal{O}_{\mathbb{P}}(H^{-n-2})$, which in turn generated canonically the corresponding sheaves $\mathcal{S}_{\mathbb{F}}(n)$ and $\mathcal{S}_{\mathbb{M}}(n)$ on $\mathbb{F}$ and $\mathbb{M}$, where sections of $\mathcal{S}_{\mathbb{M}}(n)$ turn out to be spinor fields of spin $^n/2$, for $n \geq 0$. The fibration $\mathbb{F} \xrightarrow{\alpha} \mathbb{P}$ gave rise to a cotangent bundle T^*_α along the fibers and a differential operator d_α along the fibers. Representing cohomology in terms of differential forms then gave us the diagram (3), where α^* is just the pullback of differential forms and d_α is exterior differentiation of differential forms along the fibers. Then it is intuitively clear that if $\phi \in H^1(\mathbb{P}_+, \mathcal{S}_{\mathbb{P}})$, then $d_\alpha \alpha^* \phi = 0$, since $\alpha^*\phi$ will be constant along the fibers. Then general spectral sequence theory allows us to conclude that the natural vertical mappings I and I_α (which correspond explicitly to integration over the fibers of β), are isomorphisms. Thus the induced differential operator ∇_α annihalates $\mathcal{P}\phi = I \circ \alpha^* \phi$, for $\phi \in H^1(\mathbb{P}_+, \mathcal{S}_{\mathbb{P}}(n))$. Intuitively, the pullback of the cohomology class when integrated over the fibers of β must satisfy the field equations, and all solutions arise this way (see [12] for an explicit computation of the fact that $\nabla_\alpha \mathcal{P}\phi = 0$, due to N. Woodhouse).

What is the mapping I in (3)? Given a cohomology class $\psi \in H^1(\mathbb{F}_+, \mathcal{S}_{\mathbb{F}})$, and a point $p \in \mathbb{M}_+$, ψ defines a cohomology class in any neighborhood of $\beta^{-1}(p)$, by restriction, and thus an element in $(\beta^1_* \mathcal{S}_{\mathbb{F}})_p$, the stalk of the sheaf $\mathcal{S}_{\mathbb{M}}$ at p. Loosely said, the cohomology class ψ "restricted to the fiber" gives an element in the "cohomology along the fiber," which is the desired evaluation mapping I. Notice that the fibers of β are simply

1-dimensional projective spaces, and are, in fact, submanifolds of $\mathbb{P}$. Let $\tau^{-1}(p) = Y_p \subset \mathbb{P}$, for $p \in \mathbb{M}$ (cf. (1)), and let $\mathcal{O}_{Y_p}$ be the sheaf of holomorphic functions on Y_p. Then one can show easily that the basic sheaf $\mathcal{S}_{\mathbb{M}}(n)$ is the sheaf of holomorphic sections of the holomorphic vector bundle $V_n \to \mathbb{M}$, where the fibers of the bundle $V_{n,p}$ for $p \in \mathbb{M}$ is given by

$$V_{n,p} = H^1(Y_p, \mathcal{O}_{Y_p}(H^{-n-2})), \tag{5}$$

where H^{-n-2} means $H^{-n-2}|_{Y_p}$ which is the same as the hyperplane section bundle of $\mathbb{P}_1 \cong Y_p$, raised to the (-n-2) power. We could have defined

$$V_n = \bigcup_{p \in \mathbb{M}} H^1(Y_p, \mathcal{O}_{Y_p}(H^{-n-2})) \tag{6}$$

but it is not as clear from this definition that V_n is a well-defined smooth or holomorphic vector bundle as p varies in $\mathbb{M}$, since Y_p might intersect Y_q for $p \neq q$. That is the reason for using the space $\mathbb{F}$ (which splits the fibers apart) and the direct image sheaves. But the Penrose transform says simply from this point of view: take $\phi \in H^1(\mathbb{P}_+, \mathcal{O}_{\mathbb{P}}(H^{-n-2}))$, and restrict ϕ to the submanifold $Y_p \subset \mathbb{P}_+$, for $p \in \mathbb{M}_+$, thus obtaining an element in $H^1(Y_p, \mathcal{O}_{Y_p}(H^{-n-2}))$, using, for instance, differential forms and their restrictions to submanifolds as differential forms to represent cohomology and the restriction mapping on cohomology. Here is why we assume $n \geq 0$, since in that case

$$\begin{aligned} H^1(Y_p, \mathcal{O}_{Y_p}(H^{-n-2})) &\cong H^1(\mathbb{P}_1, \mathcal{O}_{\mathbb{P}_1}(H^{-n-2})) \\ &\cong H^0(\mathbb{P}_1, \mathcal{O}(H^n)) \\ &\cong \odot^n(\mathbb{C}^2). \end{aligned}$$

Whereas if $n < 0$, we have

$$H^1(Y_p, \mathcal{O}_{Y_p}(H^{-n-2})) = 0$$

by e.g. the Kodaira vanishing theorem (cf. [10]), i.e. $\beta^1_*(S_{\mathbb{F}}(n)) = 0$, if $n < 0$, as indicated earlier.

So consider the problem of evaluating a cohomology class $\phi \in H^1(\mathbb{P}_+, S_{\mathbb{P}}(n))$ and obtaining a spinor field for $n < 0$. Simple restriction to the fibers Y_p by the above process gives zero, as the vector space where the evaluation takes place itself vanishes! Nevertheless Penrose has given contour integral formulas [7] which show how to evaluate explicit representations of such cohomology classes to obtain fields of the form (using twistor notation)

$$\phi_{AB\ldots D}(x) = \oint \frac{\partial}{\partial \omega^A} \cdots \frac{\partial}{\partial \omega^D} f(Z^A)\pi^{E'} d\pi_{E'}, \tag{7}$$

where f has homogeneity $-n-2$, and $n < 0$, and the integral is over an appropriate contour in $\tau^{-1}(x) = Y_x$.

We want to evaluate the cohomology class ϕ considered above in terms of the leading term of an expansion of the cohomology class in a power series expansion about the fiber $Y_p = \tau^{-1}(p)$, for some $p \in \mathbb{M}_+$. We will let $Y = Y_p$, and let p remain fixed in this discussion. Let I_Y be the ideal sheaf of the sections of $\mathcal{O}_{\mathbb{P}}$ which vanish on Y, and consider the short exact sequences of sheaves:

$$\begin{aligned} &0 \to I_Y \to \mathcal{O}_{\mathbb{P}} \to \mathcal{O}_{\mathbb{P}}/I_Y \to 0 \\ &0 \to I_Y^2 \to I_Y \to I_Y/I_Y^2 \to 0 \\ &\qquad\vdots \\ &0 \to I_Y^{m+1}/I_Y^m \to I_Y^m \to I_Y^m/I_Y^{m+1} \to 0. \end{aligned} \tag{8}$$

Now I_Y^m/I_Y^{m+1} can be identified with the $m^{\underline{th}}$ symmetric power of the dual to the normal bundle to the embedding $Y \subset \mathbb{P}$ (cf. Grauert [3]). We can see that if $N \to Y$ is the normal bundle to Y in $\mathbb{P}$, then $N \cong H \oplus H$, where H is the hyperplane section bundle of $\mathbb{P}$ restricted to Y (cf. Penrose [6]) and thus $N^* \cong H^{-1} \oplus H^{-1}$. It follows that

$$I_Y^{m+1}/I_Y^m \cong \mathcal{O}_Y(\odot^m N^*)$$

and

$$H^1(\mathbb{P}_+, I_Y^{m+1}/ I_Y^m) \cong H^1(Y, \mathcal{O}_Y(\odot^m N^*)) .$$

Now tensor $S_{\mathbb{P}}(n)$ (as $\mathcal{O}_{\mathbb{P}}$-modules) with the above sequences (8) obtaining

$$\begin{gathered} 0 \to I_Y \otimes S_{\mathbb{P}}(n) \to S_{\mathbb{P}}(n) \to \mathcal{O}_Y(H^{-n-2}) \to 0, \\ 0 \to I_Y^2 \otimes S_{\mathbb{P}}(n) \to I_Y \otimes S_{\mathbb{P}}(n) \to \mathcal{O}_Y(H^{-n-2} \otimes N^*) \to 0, \\ \vdots \qquad\qquad \vdots \qquad\qquad \vdots \\ 0 \to I_Y^{m+1} \otimes S_{\mathbb{P}}(n) \to I_Y^m \otimes S_{\mathbb{P}}(n) \to \mathcal{O}_Y(H^{-n-2} \otimes \odot^m N^*) \to 0. \end{gathered} \tag{9}$$

Look at the associated long exact sequences of cohomology on $\mathbb{P}_+$:

$$\begin{gathered} \ldots \to H^1(\mathbb{P}_+, I_Y \otimes S_{\mathbb{P}}(n)) \xrightarrow{i_0} H^1(\mathbb{P}_+, S_{\mathbb{P}}(n)) \xrightarrow{r_0} H^1(Y, \mathcal{O}_Y(H^{-n-2})) \to \ldots, \\ \ldots \to H^1(\mathbb{P}_+, I_Y^2 \otimes S_{\mathbb{P}}(n)) \xrightarrow{i_1} H^1(\mathbb{P}_+, I_Y \otimes S_{\mathbb{P}}(n)) \to \\ \xrightarrow{r_1} H^1(Y, \mathcal{O}_Y(H^{-n-2} \otimes N^*)) \to \ldots, \\ \vdots \qquad\qquad \vdots \\ \ldots \to H^1(\mathbb{P}_+, I_Y^{m+1} \otimes S_{\mathbb{P}}(n)) \xrightarrow{i_m} H^1(\mathbb{P}_+, I_Y^m \otimes S_{\mathbb{P}}(n)) \to \\ \xrightarrow{r_m} H^1(Y, \mathcal{O}_Y(H^{-n-2} \otimes \odot^m N^*)) \to \ldots . \end{gathered} \tag{10}$$

Now we note that $Y \cong \mathbb{P}_1(\mathbb{C})$ and if we now assume that $n < 0$, then by Kodaira's vanishing theorem

$$H^1(Y, \mathcal{O}_Y(H^{-n-2} \otimes \odot^m N^*)) = 0$$

for $m < -n$. Recalling that $Y = Y_p$, for $p \in \mathbb{M}_+$ in the above arguments and that $N^* \cong H^{-1} \oplus H^{-1}$, we find that if we let, for $n < 0$,

$$V_{n,p} = H^1(Y_p, \mathcal{O}_{Y_p}(H^{-n-2} \otimes \odot^{-n}(H^{-1} \oplus H^{-1}))), \tag{11}$$

then we find that

$$V_{n,p} = H^1(Y_p, \mathcal{O}_{Y_p}(H^{-2})) \otimes_{\mathbb{C}} \odot^{-n}\mathbb{C}^2 \cong \odot^{-n}\mathbb{C}^2.$$

Thus the vector space $V_{n,p}$ is a vector space of spinor type, for each $p \in \mathbb{M}_+$ (although it is clear that to define $V_{n,p}$ by (11) we can have $p \in \mathbb{M}$, not just $p \in \mathbb{M}_+$).

By defining V_n, for $n < 0$, by (11) to be the "holomorphic vector bundle with fiber $V_{n,p}$," we see that we can evaluate a cohomology class $\phi \in H^1(\mathbb{P}_+, \mathcal{S}_{\mathbb{P}}(n))$, for $n < 0$, on Y_p to give an element of $V_{n,p}$ for each $p \in \mathbb{M}_+$. We will see later that V_n so defined is indeed a bundle, but we will discuss the evaluation procedure first. If $\phi \in H^1(\mathbb{P}_+, \mathcal{S}_{\mathbb{P}}(n))$ is given, for some $n < 0$, then we see that in (10), $r_0(\phi) = 0$, so $\phi = i_0(\phi_1)$, with $\phi_1 \in H^1(\mathbb{P}_+, \mathcal{I}_{Y_p} \otimes \mathcal{S}_{\mathbb{P}}(n))$. If $n = -1$, then $r_1(\phi_1) \in V_{-1,p}$, and we have evaluated ϕ on Y_p. If $n < -1$, then $r_1(\phi_1) = 0$, and $\phi_1 = i_1(\phi_2)$, and we consider $r_2(\phi_2)$, etc. Finally, by induction, we arrive at $r_{-n}(\phi_{-n}) \in V_{n,p}$. One checks that the element of $V_{n,p}$ obtained in this manner is independent of the choices made, and that a canonical evaluation of ϕ on Y_p has been made. This corresponds to the $(-n)$-th term of the Taylor expansion of ϕ about the fiber Y_p, and is the leading term in such an expansion (cf. Grauert [3]). It depends on the normal bundle information up to the (-n)-th order, just as does the formula of Penrose [7], but it does so in an intrinsic and equivariant manner. The only remaining problem is to understand why $\bigcup_{p \in \mathbb{M}} V_{n,p}$, for $n < 0$, is indeed a holomorphic vector bundle over $\mathbb{M}$. To do this we need to carry out the above evaluation process, and hence the ideal-theoretic arguments in a uniform and smooth manner, not simply one fiber at a time. The basic idea is to consider

$$\mathbb{P}_+ \times \mathbb{M}_+$$

and

$$\mathcal{Y} \subset \mathbb{P}_+ \times \mathbb{M}_+$$

defined by

$$\mathcal{Y} = \bigcup_{p \in \mathbb{M}_+} (Y_p \times \{p\}),$$

a complex submanifold of complex codimension 2 of $\mathbb{P}_+ \times \mathbb{M}_+$ with the property that $\mathcal{Y} \cong \mathbb{F}_+$. One sees readily that $\mathcal{Y} \subset \mathbb{P}_+ \times \mathbb{M}_+$ has a normal bundle $N \to \mathcal{Y}$ with the property that $N|_{Y_p} \to Y_p$ is the normal bundle of Y_p in $\mathbb{P}$. In fact N is given by $\pi^*(H \oplus H)|_{\mathcal{Y}}$, where $\pi: \mathbb{P}_+ \times \mathbb{M}_+ \to \mathbb{P}_+$ is the natural product projection, and $H \to \mathbb{P}$ is the hyperplane section bundle of $\mathbb{P}$, as before. We now consider $\pi^*(\mathcal{S}_{\mathbb{P}}(n))$ on $\mathbb{P}_+ \times \mathbb{M}_+$, and let $\mathcal{I}_{\mathcal{Y}} \subset \mathcal{O}_{\mathbb{P}_+ \times \mathbb{M}_+}$ be the ideal sheaf of $\mathcal{Y}$ in $\mathbb{P}_+ \times \mathbb{M}_+$. In (9) and (10) we replace $\mathcal{I}_Y$ by $\mathcal{I}_{\mathcal{Y}}$ and $\mathcal{S}_{\mathbb{P}}(n)$ by $\pi^*(\mathcal{S}_{\mathbb{P}}(n))$ obtaining analogous short and long exact sequences on $\mathbb{P}_+ \times \mathbb{M}_+$. It follows from the same kind of vanishing theorem considerations on $\mathbb{M}_+ \times \mathbb{P}_+$ that we have the following diagram:

$$
\begin{array}{ccc}
H^1(\mathcal{Y}, \mathcal{O}_{\mathcal{Y}}(\pi^*(H^{-n-2} \otimes \odot^{-n} N_0^*))) & \cong & H^1(\mathbb{F}_+, \mathcal{O}_{\mathbb{F}}(\alpha^*(H^{-n-2} \otimes \odot^{-n} N_0^*))) \\
\Big\uparrow T & & \cong \Big\downarrow C \\
H^1(\mathbb{P}_+ \times \mathbb{M}_+, \pi^*(S_{\mathbb{P}}(n))) & & H^1(\mathbb{F}_+, \mathcal{O}_{\mathbb{F}}(\alpha^* H^{-2})) \otimes_{\mathbb{C}} \odot^{-n} \mathbb{C}^2 \\
\Big\uparrow \pi^* & & \cong \Big\downarrow I \\
H^1(\mathbb{P}_+, S_{\mathbb{P}}(n)) & \xrightarrow{\;\mathcal{P}\;} & H^0(\mathbb{M}_+, \beta^1_*(\alpha^* H^{-2})) \otimes_{\mathbb{C}} \odot^{-n} \mathbb{C}^2 .
\end{array}
\tag{12}
$$

Here T is the Taylor expansion evaluation using the analogues to (9) to (10), and where $N_0^* = H^{-1} \oplus H^{-1}$ is a bundle over $\mathbb{P}$, whose pullback $N^* = \pi^* N_0^*$ is the conormal bundle on $\mathcal{Y}$. The top horizontal isomorphism comes from a comparison of the pullbacks by α^* and π^* respectively, the mapping C is the natural contraction obtained by tensoring H with H^{-1}, and powers of same, and I is our direct image map (integration over the fiber again). The resulting mapping $\mathcal{P}$ we call the <u>Penrose transform for negative helicity</u>. (i.e. for $n < 0$). If we define V_n, for $n < 0$, to be the holomorphic vector bundle on $\mathbb{M}$ such that

$$\mathcal{O}_{\mathbb{M}}(V_n) = \beta^1_*(\alpha^* H^{-2}) \otimes_{\mathbb{C}} \odot^{-n} \mathbb{C}^2,$$

then we see that

$$\mathcal{P}: H^1(\mathbb{P}_+, S_{\mathbb{P}}(n)) \to H^0(\mathbb{M}_+, \mathcal{O}_{\mathbb{M}}(V_n)) \tag{13}$$

is a well-defined mapping and that

$$V_{n,p} = H^1(Y_p, \mathcal{O}_{Y_p}(H^{-2})) \otimes_{\mathbb{C}} \odot^{-n} \mathbb{C}^2$$

as desired. Moreover, the sections of V_n can be identified with spinor fields of unprimed type, i.e. spinors fields of the form $\{\phi_{\underbrace{AB\ldots D}_{-n}}\}$ on open subsets of $\mathbb{M}$.

This is thus the Penrose transform p extended to negative helicity. One must extend the differential operator d_α to this situation obtaining mappings from the various spaces in (12) and finally that the image of p satisfies the zero-rest-mass field equations. This is not difficult, but will not be carried out here (cf. [1]). Note also that we have neglected the case $n = 0$. This is a borderline case, and the Penrose transform p is well-defined with only the zero$^{\underline{th}}$ order Taylor expansion, but the image of p satisfies a second order differential equation (the scalar wave equation), but this is not readily apparent from the above analysis. This case is considered in detail in [1].

Finally we want to indicate how one can extend the Penrose transform to obtain solutions of the field equations on M which are not smooth. The basic result is that hyperfunction cohomology on $P \subset \mathbb{P}$ is transformed onto all hyperfunction solutions of the zero-rest-mass field equations on $M \subset \mathbb{M}$. To do this we need to use relative cohomology and understand something about taking boundary values of cohomology classes defined in $\mathbb{P}_\pm$ on the common boundary P. First we want to see how we can obtain real-analytic solutions of the field equations on M. We will consider the case only where the helicity is positive, since the analysis is simplest there, as we see from the previous discussions.

Let $\mathbb{M}_U$ be an open neighborhood of M in $\mathbb{M}$, and let $\mathbb{F}_U = \beta^{-1}(\mathbb{M}_U)$, and $\mathbb{P}_U = \alpha(\beta^{-1}(\mathbb{M}_U))$. Thus we have our basic diagram

$$\begin{array}{ccc} & \mathbb{F}_U & \\ {}^{\alpha}\swarrow & & \searrow^{\beta} \\ \mathbb{P}_U & & \mathbb{M}_U \end{array}$$

with our basic sheaves $\mathcal{S}_{\mathbb{P}}(n)$, $\mathcal{S}_{\mathbb{F}}(n)$, and $\mathcal{S}_{\mathbb{M}}(n)$ defined on these open sets. For simplicity we will omit the notational dependence on n, assumed positive. We have the Penrose transform defined by

$$\begin{array}{ccc} & H^1(\mathbb{F}_U, \mathcal{S}_{\mathbb{F}}) & \\ {}^{\alpha^*}\nearrow & & \searrow^{I} \\ H^1(\mathbb{P}_U, \mathcal{S}_{\mathbb{P}}) & \xrightarrow{\ p\ } & H^0(\mathbb{M}_U, \mathcal{S}_{\mathbb{M}}). \end{array}$$

Now if U is a fundamental system of neighborhoods of M we see that p expends to the direct limit

$$p\colon \varinjlim_{\mathbb{P}_U \supset P} H^1(\mathbb{P}_U, \mathcal{S}_{\mathbb{P}}) \longrightarrow \varinjlim_{\mathbb{M}_U \supset M} H^0(\mathbb{M}_U, \mathcal{S}_{\mathbb{M}}). \tag{14}$$

We see that the direct limit on the right is, by definition, real-analytic spinor fields on M. It is not clear from this analysis that the real-analytic solutions of the massless field equations are precisely the image of p in (14), but this is nevertheless the case (see [13]). Let us denote

$$\begin{array}{ccc} \varinjlim_{\mathbb{P}_U \supset P} H^1(\mathbb{P}_U, \mathcal{S}_{\mathbb{P}}) = H^1_A(P, \mathcal{S}_{\mathbb{P}}) & & \\ \swarrow^{\alpha^*} & & \Big\downarrow p \\ \varinjlim_{\mathbb{F}_U \supset F} H^1(\mathbb{F}_U, \mathcal{S}_{\mathbb{F}}) = H^1_A(F, \mathcal{S}_{\mathbb{F}}) & & \\ \searrow^{I} & & \\ \varinjlim_{\mathbb{M}_U \supset M} H^0(\mathbb{M}_U, \mathcal{S}_{\mathbb{M}}) = H^0_A(M, \mathcal{S}_{\mathbb{M}}), & & \end{array} \tag{15}$$

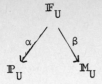

the subscript $\mathcal{A}$ denoting real-analytic. Thus we have $\mathcal{P}$ defined on real-analytic data on P.

If $S \subset X$ is a closed subset of a topological space X, and $\mathcal{F}$ is a sheaf of abelian groups on X, then we will let $H^q_S(X,\mathcal{F})$ denote the relative cohomology of X with coefficients in $\mathcal{F}$ relative to the subset X. This is also often referred to as cohomology with supports in S. There is a long exact sequence of relative cohomology

$$\cdots \to H^q(X,\mathcal{F}) \to H^q(X-S,\mathcal{F}) \to H^{q+1}_S(X,\mathcal{F}) \to \cdots$$

which shows that the relative cohomology groups are the obstructions that a cohomology class on X - S be the restriction of a cohomology class defined on all of X (cf. Schapira [9], for instance, for a discussion of this as well as of hyperfunctions, which we will use below.) Hyperfunctions on M can be defined as

$$\mathcal{B}(M) = H^4_M(\mathbb{M}, \mathcal{O}_{\mathbb{M}}),$$

and it is a deep result of Sato that this is the same as the dual of the real-analytic functions $\mathcal{A}(M)'$, where $\mathcal{A}(M)$ is equipped with the inductive limit topology of Frechet spaces of holomorphic functions,

$$\mathcal{A}(M) = \varinjlim_{\mathbb{M}_U \supset M} \mathcal{O}(\mathbb{M}_U).$$

Knowing that $\mathcal{B}(M) = \mathcal{A}(M)'$, it is then easy to see that $\mathcal{B}(M)$ contains in a natural manner the distributions on M, $\mathcal{D}'(M)$, since $\mathcal{A}(M)$ is dense in $\mathcal{D}(M)$. But the interpretation of hyperfunctions as relative cohomology gives a powerful means of working with hyperfunctions (cf. Schapira [9]). We will use this representation of hyperfunctions to indicate how the Penrose transform can be extended to the level of hyperfunctions (cf. the extension of the Fourier transform from smooth functions to distributions of the appropriate type). First we need some appropriate "hyperfunction cohomology" on P. We consider

$$\to H^1(\mathbb{P}, \mathcal{S}_{\mathbb{P}}) \to H^1(\mathbb{P} - P, \mathcal{S}_{\mathbb{P}}) \to H^2_P(\mathbb{P}, \mathcal{S}_{\mathbb{P}}) \to H^2(\mathbb{P}, \mathcal{S}_{\mathbb{P}}) \to \dots,$$

and one knows that $H^1(\mathbb{P}, \mathcal{S}_{\mathbb{P}}) = H^2(\mathbb{P}, \mathcal{S}_{\mathbb{P}}) = 0$, and thus

$$H^2_P(\mathbb{P}, \mathcal{S}_{\mathbb{P}}) \cong H^1(\mathbb{P} - P, \mathcal{S}_{\mathbb{P}}) \cong H^1(\mathbb{P}_+, \mathcal{S}_{\mathbb{P}}) \oplus H^1(\mathbb{P}_-, \mathcal{S}_{\mathbb{P}}).$$

Intuitively, $H^2_P(\mathbb{P}, \mathcal{S}_{\mathbb{P}})$ corresponds to the "jump" of the "boundary values" of $H^1(\mathbb{P}_+, \mathcal{S}_{\mathbb{P}})$ and $H^1(\mathbb{P}_-, \mathcal{S}_{\mathbb{P}})$ on P, and this can be given a precise interpretation. Ordinary hyperfunctions on $\mathbb{R}^n$ can also be considered as jumps of boundary values of holomorphic functions defined in open subsets of $\mathbb{C}^n - \mathbb{R}^n$ (cf. Schapira [9]), and we have the same phenomenon at the level of the first cohomology groups. In particular, one can interpret $H^2_P(\mathbb{P}, \mathcal{S}_{\mathbb{P}})$ as intrinsic hyperfunction cohomology on P determined by the $\bar{\partial}_P$-complex on P acting on hyperfunction-valued (0,2)-forms. See [8] for a discussion of these points.

If $Y \subset X$ is a real submanifold of a complex manifold X of real codimension r, then [Y] is the current of integration over Y. It is a current (differential form with distribution coefficients) of degree r (cf. Harvey [4]). Since X is a complex manifold, any current ϕ of degree r can be written uniquely as the sum of currents of specified bidegree or type

$$\phi = \phi^{r,0} + \phi^{r-1,0} + \dots + \phi^{0,r} ,$$

just as is the case for differential forms with smooth coefficients. So in particular, if Y is the submanifold above of codimension r, then

$$[Y] = [Y]^{r,0} + \ldots + [Y]^{0,r} ,$$

and $[Y]^{0,r}$ is a singular current of type (0,r) with support on Y.

We now apply the above analysis to our submanifolds $P \subset \mathbb{P}$, etc., to obtain an embedding of real-analytic cohomology into hyperfunction cohomology as follows:

$$\begin{array}{ccc} H^1(P, \mathcal{S}_{\mathbb{P}}) & \longrightarrow & H^2_P(\mathbb{P}, \mathcal{S}_{\mathbb{P}}), \\ \Psi & & \Psi \\ \phi & \longmapsto & \phi \wedge [P]^{0,1} \\ & & \\ H^1_A(F, \mathcal{S}_{\mathbb{F}}) & \longrightarrow & H^5_F(\mathbb{F}, \mathcal{S}_{\mathbb{F}}), \\ \Psi & & \Psi \\ \phi & \longmapsto & \phi \wedge [F]^{0,4} \\ & & \\ H^0_A(M, \mathcal{S}_{\mathbb{M}}) & \longrightarrow & H^4_M(\mathbb{M}, \mathcal{S}_{\mathbb{M}}) . \\ \Psi & & \Psi \\ \phi & \longmapsto & \phi \wedge [M]^{0,4} \end{array} \tag{16}$$

We see that, for instance in the first case, $\phi \wedge [P]^{0,1}$ is indeed a $\bar{\partial}$-closed (0,2)-current on $\mathbb{P}$ with support in P, and will represent a class in $H^2_P(\mathbb{P}, \mathcal{S}_{\mathbb{P}})$, and similarly for the other cases. Here we are representing $H^2_P(\mathbb{P}, \mathcal{S}_{\mathbb{P}})$, as $\bar{\partial}$-closed (0,2)-hyperfunction forms on $\mathbb{P}$ with coefficients in $\mathcal{S}_{\mathbb{P}}$, modulo $\bar{\partial}$-exact forms of the same type. This is a generalization of the usual Dolbeault isomorphism. It would be impossible to represent relative cohomology (or cohomology with support in P) in terms of smooth differential forms. A hyperfunction form here is a generalization of a current, allowing hyperfunction coefficients instead of distribution coefficients, so a current such as $[P]^{0,1}$ is certainly a hyperfunction form of type (0,1).

We will have extended the Penrose transform p to the hyperfunction level if we can extend the mappings α^* and I in (15) to the right hand column in

(16) to be compatible with the horizontal injections in (16). This is possible, and the principal tool is to introduce a "current of integration along the fiber of α" to be denoted by $[\alpha]^{0,3}$, which is defined in the following manner. Consider the pair of fibrations:

$$\begin{array}{ccc} F & \longrightarrow & \mathbb{F} \\ \Big\downarrow{\scriptstyle \alpha|_F} & & \Big\downarrow{\scriptstyle \alpha} \\ P & \longrightarrow & \mathbb{P} \end{array}$$

the left hand fibration $F \to P$ has fiber S^1, and the right hand fibration has fiber $\mathbb{P}_2(\mathbb{C})$, and we denote the inclusion of fibers by $F_p \subset \mathbb{F}_p$, for $p \in P$. Thus F_p is a 3-codimensional submanifold of $\mathbb{F}_p$ and as such defines a current of integration on $\mathbb{F}_p$ denoted by $[F_p]$, which has a component $[F_p]^{0,3}$. This is a family of currents parametrized by $p \in F$. We denote this F-parameter family of currents symbolically by $[\alpha]^{0,3}$, and one shows that one can form the tensor product of this measure along the fibers with currents or hyperfunction forms representing elements of $H^2_p(\mathbb{P}, S_{\mathbb{P}})$ to define hyperfunction forms on $\mathbb{F}$ of type (0,5). This is similar to constructing product measures on Cartesian products of measure spaces. This is the required "pullback" of the cohomology class in $H^2_P(\mathbb{P}, S_{\mathbb{P}})$ to $H^5_P(\mathbb{F}, S_{\mathbb{F}})$, and we denote this mapping by $\tilde{\alpha}^*$. One can also extend the Leray spectral sequence argument and direct image mapping I to the case of cohomology with supports in F and M, and extend the mapping I to a mapping $\tilde{I}$ giving us then the following commutative diagram:

$$\begin{array}{ccccc} & H^1_A(P,S_{\mathbb{P}}) & \longrightarrow & H^2_P(\mathbb{P},S_{\mathbb{P}}) & \\ \alpha^* \swarrow & & \tilde{\alpha}^* \swarrow & & \downarrow p \\ H^1(F,S_{\mathbb{F}}) & \longrightarrow & H^5_F(\mathbb{F},S_{\mathbb{F}}) & & \\ & \searrow I & & \tilde{I} \searrow & \\ & H^0(M,S_{\mathbb{M}}) & \longrightarrow & H^4_M(\mathbb{M},S_{\mathbb{M}}) & \end{array}$$

and we have defined $p = \tilde{I} \circ \tilde{\alpha}^*$. It is again necessary to extend the differential operator d_α to this setting which is not difficult, and one obtains the result that

$$p : H^2_P(\mathbb{P}, S_{\mathbb{P}}) \rightarrow H^4_M(\mathbb{M}, S_{\mathbb{M}})$$

maps holomorphic hyperfunction data on P to hyperfunction solutions of the field equations on M (in positive helicity in the case above), and moreover, all hyperfunction solutions on M arise in this manner (see [13] for the proof of this theorem).

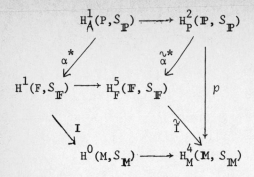

There are numerous problems which remain in this area, e.g. extending the Penrose transform to coupled interactions of zero-rest-mass fields and Yang-Mills fields, or to the more general geometric situation involving deformations of subsets of $\mathbb{P}$ which arises in Penrose's nonlinear graviton [6]. There are various versions of inverse transforms to the Penrose transform, which have been developed by Hughston, Penrose, and Ward (cf. various issues of the Twistor Newsletter from Oxford), and such inverses will be discussed in more detail in [1]. We will close with one philosophical remark. The power of the Fourier transform is that it converts problems concerning differential operators to problems involving algebraic operators. The Penrose transform has this feature also. Namely the differential operators of mathematical physics get replaced by the algebraic operators involving sheaf cohomology. It would be worthwhile to understand this phenomenon independent of the particular differential equations that one wants to study, as has been the case here, and to see how close an analogue of the powerful Fourier transform this new transform of Penrose's really is.

References

1 M. Eastwood, R. Penrose, and R. O. Wells, Jr., "Cohomology and massless fields," (to appear).

2 Roger Godement, Topologie Algébrique et Théorie des Faisceau, Hermann & Cie, Paris, 1964.

3 H. Grauert, "Über Modifikationen and exzeptionelle analytische Mengen" Math. Ann. 146(1962), 331-368.

4 R. Harvey, "Holomorphic chains and their boundaries," Proc. Symp. Pure Math. Vol. 30, Part 1, Amer. Math. Soc., Providence, R.I., 1977, 309-382.

5 R. Penrose, "Twistor algebra," J. Math. Phys., 8(1967), 345-366.

6 R. Penrose, "Nonlinear gravitons and curved twistor theory," General Relativity and Gravitation, 7(1976), 31-52.

7 R. Penrose, "The twistor program" Rep. on Math. Phys. 12, 65(1977).

8 J. Polking and R. O. Wells, Jr., "Boundary values of Dolbeault cohomology classes and generalized Bochner-Hartogs Theorem," Abhand. Math. Sem. Univ. Hamburg, 47(1978), 1-24.

9 P. Schapira, Theorie des Hyperfunctions, Lecture Notes in Math., Vol. 126, Springer-Verlag, Berlin-Heidelberg-New York, 1970.

10 R. O. Wells, Jr., Differential Analysis on Complex Manifolds, Prentice-Hall, Inc. Englewood Cliffs, N.J., 1973.

11 R. O. Wells, Jr., "Function theory on differentiable submanifolds," Contributions to Analysis, Academic Press, 1974, 407-441.

12 R. O. Wells, Jr., "Complex manifolds and mathematical physics," Bull. Amer. Math. Soc. (to appear).

13 R. O. Wells, Jr., "Hyperfunction solutions of the zero-rest-mass field equations," (to appear).

Acknowledgment

The research is supported by NSF MCS 78-03571.

R O WELLS, JR
Department of Mathematics
Rice University
Houston, TX 77001

L P Hughston

Some new contour integral formulae

1. INTRODUCTION

The natural setting[(1)] for twistor theory — and in particular, those aspects of the theory concerned primarily with zero rest mass phenomena — is complex projective three-space P^3. In connection with the problem of understanding the role of massive fields in the theory it is of great interest that much of the analysis generalizes, in a remarkably straightforward way, when one goes up from P^3 to P^{2n+1}, where n is any positive integer. The basic set-up is as follows.

Let S^a denote C^{n+1} regarded as a complex vector space; and denote by S_a, $S^{a'}$, and $S_{a'}$ respectively the dual space, the complex conjugate space, and the dual complex conjugate space to S^a. By a *generalized twistor* we mean a point in the space $(S^a, S_{a'})$, i.e. a pair $(\omega^a, \pi_{a'})$ with $\omega^a \in S^a$ and $\pi_{a'} \in S_{a'}$. Let us denote the space $(S^a, S_{a'})$ by T^α; in the event that $n = 1$, T^α is clearly twistor space. In the general case we have $T^\alpha \cong C^{2n+2}$, and the associated projective space in P^{2n+1}.

The dual space to T^α is $T_\alpha = (S_a, S^{a'})$. For any generalized twistor $Z^\alpha = (\omega^a, \pi_{a'})$ we define its complex conjugate by $\overline{Z}_\alpha = (\overline{\pi}_a, \overline{\omega}^{a'})$. Clearly, we have $\overline{Z}_\alpha \in T_\alpha$. If $W_\alpha = (\sigma_a, \tau^{a'})$ is a dual twistor then the inner product $Z^\alpha W_\alpha$ is defined to be $\omega^a \sigma_a + \pi_{a'} \tau^{a'}$. In particular, the norm of Z^α is defined to be $Z^\alpha \overline{Z}_\alpha$. This norm has signature $(n+1, n+1)$. We shall denote by T^+, T^-, and N those portions of T^α for which the norm is positive, negative, and zero, respectively; the associated regions of the projective space PT will be denoted PT^+, PT^-, and PN, respectively.

The Grassmann variety of projective n-planes in P^{2n+1} has dimension $(n+1)^2$. In the case n = 1 the Grassmann variety can be regarded — and this fact is perhaps the true starting point for modern applications of algebraic geometry to problems in theoretical physics — as complex Minkowski space, with some structure added in at infinity(2). A similar interpretation is valid for general n, and the "finite" points of the Grassmannian in that case can be represented by points in the complex vector space $S^a \otimes S^{a'}$, a typical point being denoted $x^{aa'}$. A twistor $Z^\alpha = (\omega^a, \pi_{a'})$ lies on the projective n-plane determined by $x^{aa'}$ if and only if $\omega^a = ix^{aa'}\pi_{a'}$.

Let us denote the above-mentioned Grassmannian by M. It is noteworthy that certain categories of field equations can be introduced on M which — in an entirely natural and reasonable way — generalize analogous field equations on Minkowski space. Suppose one considers, for example, the neutrino equation

$$\nabla_{AB'}\phi^{B'} = 0 \quad , \tag{1.1}$$

where $\phi^{B'}(x^{AA'})$ is the neutrino wave function. How can equation (1.1) be suitably generalized to arbitrary values of n? First, we rewrite (1.1) in the form $\nabla_{A[A'}\phi_{B']} = 0$. Then, we generalize by making the following substitutions:

$$\nabla_{AA'} \to \nabla_{aa'} \quad , \quad \phi_{B'}(x^{AA'}) \to \phi_b(x^{aa'}) \quad , \tag{1.2}$$

where $\nabla_{aa'} = \partial/\partial x^{aa'}$. When these substitutions are made in (1.1) the result is

$$\nabla_{a[a'}\phi_{b']} = 0 \quad , \tag{1.3}$$

which one can tentatively regard as the proper generalization of (1.1). It should be noted moreover that equation (1.1) implies automatically the wave equation $\Box\phi_{B'} = 0$. The wave equation can be written in the form $\nabla_{A[A'}\nabla_{B']B}\phi_{C'} = 0$, and thus generalizes, using (1.2), to:

$$\nabla_{a[a'}\nabla_{b']b}\phi_{c'} = 0 \quad . \tag{1.4}$$

For $n \neq 1$ equation (1.3) likewise gives equation (1.4), and therefore — for reasons to be justified in Section 3 — we can regard (1.3) and (1.4) *together* as comprising the correct generalization of the neutrino equation.

In what follows certain aspects of the geometry of higher dimensional twistor-type spaces will be discussed, and in particular some new contour integral formulae will be introduced — modelled after the extraordinary twistor contour integral formulae that Penrose used [3] for solving the zero rest mass equations in Minkowski space — in order to solve the combined system of equations (1.3) and (1.4), and also to treat certain related systems of equations.

2. A GENERALIZATION OF THE TWISTOR EQUATION

From a Minkowski space viewpoint projective twistors can be characterized as solutions of the differential equation $\nabla^{A'(A}\xi^{B)} = 0$. Indeed, the solution of this equation[3] is $\xi^A = \omega^A - ix^{AA'}\pi_{A'}$, and the associated twistor is determined by the fixed spinor pair $(\omega^A, \pi_{A'})$. Now consider the equation

$$(n+1)\,\nabla_{aa'}\xi^c = \delta^c_a\,\nabla_{da'}\,\xi^d \quad . \tag{2.1}$$

It is not difficult to check that in the case $n = 1$ equation (2.1) reduces to the twistor equation. The following result establishes the connection between solutions of (2.1) and the generalized twistors introduced in the previous section:

Theorem 1 *The general solution of equation (2.1) is given by*

$$\xi^{a} = \omega^{a} - ix^{aa'}\pi_{a'} , \tag{2.2}$$

where ω^{a} *and* $\pi_{a'}$ *are constant*[4].

Proof Differentiating (2.1) one has

$$(n+1)\nabla_{bb'}\nabla_{aa'}\xi^{c} = \delta^{c}_{a}\nabla_{bb'}\nabla_{da'}\xi^{d} , \tag{2.3}$$

which, upon interchange of the index clumps aa' and bb', can be written as

$$(n+1)\nabla_{aa'}\nabla_{bb'}\xi^{c} = \delta^{c}_{b}\nabla_{aa'}\nabla_{db'}\xi^{d} . \tag{2.4}$$

Since $\nabla_{aa'}$ and $\nabla_{bb'}$ commute, the left-hand sides of equations (2.3) and (2.4) are equal; therefore:

$$\delta^{c}_{a}\nabla_{bb'}\nabla_{da'}\xi^{d} = \delta^{c}_{b}\nabla_{aa'}\nabla_{db'}\xi^{d} . \tag{2.5}$$

Transvecting equation (2.5) with δ^{b}_{c} gives

$$\nabla_{ab'}\nabla_{da'}\xi^{d} = (n+1)\nabla_{aa'}\nabla_{db'}\xi^{d} . \tag{2.6}$$

On the other hand, transvecting (2.5) with δ^{a}_{c}, one obtains $(n+1)\nabla_{bb'}\nabla_{da'}\xi^{d} = \nabla_{ba'}\nabla_{db'}\xi^{d}$ which, substituting b → a, then reads:

$$(n+1)\nabla_{ab'}\nabla_{da'}\xi^{d} = \nabla_{aa'}\nabla_{db'}\xi^{d} . \tag{2.7}$$

Taken together, equations (2.6) and (2.7) imply

$$\nabla_{aa'}\nabla_{db'}\xi^{d} = 0 , \tag{2.8}$$

showing that $\nabla_{db'}\xi^{d}$ is constant; it follows, from (2.1), that ξ^{a} is linear in $x^{aa'}$. Inserting the most general linear expression for ξ^{a} into (2.1), a short calculation verifies that (2.2) is, as desired, the general solution. □

It should be observed that points in the dual space T_α can be represented — up to proportionality — by solutions of the equation

$$(n+1)\,\nabla_{aa'}\eta^{c'} = \delta^{c'}_{a'}\,\nabla_{ad'}\eta^{d'}\ , \tag{2.9}$$

which is the primed analogue of (2.1); following the pattern of *Theorem 1* it can be shown that the general solution of (2.9) is of the form

$$\eta^{a'} = \tau^{a'} + ix^{a'a}\sigma_a\ , \tag{2.10}$$

where the pair $(\sigma_a, \tau^{a'})$ corresponds to a point W_α in the dual space T_α.

3. CONTOUR INTEGRAL FORMULAE

Let us denote by ρ_x the operation of restriction to the projective n-plane in PT corresponding to the point $x^{aa'}$ in M. Thus for any twistor $Z^\alpha = (\omega^a, \pi_{a'})$ we write:

$$\rho_x\ Z^\alpha = (ix^{aa'}\pi_{a'}, \pi_{a'})\ . \tag{3.1}$$

Similarly, if $f(Z^\alpha)$ is a function of Z^α, then we shall write $\rho_x f(Z^\alpha) = f(\rho_x Z^\alpha)$.

Now consider the contour integral expression[5]

$$\phi_{a'}(x) = \oint \rho_x \pi_{a'}\ \psi(Z^\alpha)\Delta\pi\ , \tag{3.2}$$

where $\psi(Z^\alpha)$ is a holomorphic function — the singularity structure of which will be indicated, by means of some examples, in the next section — taken to be homogeneous of degree -n-2 in Z^α. The differential form $\Delta\pi$ is defined by

$$\Delta\pi = \varepsilon^{a'b'\ldots c'}\ \pi_{a'}d\pi_{b'}\ldots d\pi_{c'}\ . \tag{3.3}$$

Note that $\Delta\pi$ is homogeneous of degree n+1 in $\pi_{a'}$, and thus that the quantity appearing under the integral sign in (3.2) is — taken in its entirety — homogeneous of degree zero.

Theorem 2 *The field $\phi_{a'}(x)$ defined in formula (3.2) satisfies equations (1.3) and (1.4).*

Proof Using the differential chain rule it is straightforward to establish the relations

$$\begin{aligned} \nabla_{bb'} \rho_x \psi(Z^\alpha) &= \nabla_{bb'} \psi(ix^{aa'}\pi_{a'}, \pi_{a'}) \\ &= -i[\nabla_{bb'} x^{aa'}\pi_{a'}] \rho_x \hat{\pi}_b \psi(Z^\alpha) \\ &= -i \rho_x \pi_{b'} \hat{\pi}_b \psi(Z^\alpha) , \end{aligned} \tag{3.4}$$

where $\hat{\pi}_b = -\partial/\partial\omega^b$, the minus sign being introduced for later convenience. Equations (3.2) and (3.4) together readily imply the field equations (1.3) and (1.4). □

More generally, one can consider holomorphic functions homogeneous of degree -n-1-r, where r is any integer. When $r > 1$, a formula similar to (3.2) applies, only the number of π-coefficients is r, and the resulting symmetric field $\phi_{a'b'\ldots c'}(x)$, which has r indices, satisfies:

$$\left.\begin{aligned} \nabla_{d[d'} \phi_{a']b'\ldots c'} &= 0 \\ \nabla_{d[d'} \nabla_{e']e} \phi_{a'b'\ldots c'} &= 0 \end{aligned}\right\} . \tag{3.5}$$

For $r = 0$, no π-coefficients appear, and the resultant scalar field $\phi(x)$ satisfies:

$$\nabla_{a[a'} \nabla_{b']b} \phi = 0. \tag{3.6}$$

Finally, for $r < 0$ we find solutions of the equations

$$\left.\begin{aligned} \nabla_{d'[d} \phi_{a]b\ldots c} &= 0 \\ \nabla_{d[d'} \nabla_{e']e} \phi_{ab\ldots c} &= 0 \end{aligned}\right\} , \tag{3.7}$$

where the symmetric field $\phi_{ab\ldots c}$ is defined by

$$\phi_{ab\ldots c} = \oint \rho_x \hat{\pi}_a \hat{\pi}_b \ldots \hat{\pi}_c \, \psi(Z^\alpha) \Delta\pi \quad , \tag{3.8}$$

the operator $\hat{\pi}_a$ appearing $-r$ times in the integrand.

4. EXAMPLES

We shall conclude with an illustration of the procedures outlined above. It is straightforward to construct, for each value of n, fields which generalize the notion of an *elementary state*(6). Let us first take the simplest case, given by n = 2 and r = 0. For our twistor function we take

$$\psi(Z^\alpha) = [P_\alpha Z^\alpha Q_\beta Z^\beta R_\gamma Z^\gamma]^{-1} \quad , \tag{4.1}$$

where P_α, Q_β, and R_γ denote fixed points in the dual space. For their "spinor parts" we shall write

$$\left.\begin{aligned} P_\alpha &= (P_a, P^{a'}) \\ Q_\alpha &= (Q_a, Q^{a'}) \\ R_\alpha &= (R_a, R^{a'}) \end{aligned}\right\} \quad , \tag{4.2}$$

and thus, using definition (3.1), we can put

$$\left.\begin{aligned} \rho_x P_\alpha Z^\alpha &= p^{a'} \pi_{a'} \quad , \quad p^{a'} = P^{a'} + ix^{a'a} P_a \\ \rho_x Q_\beta Z^\beta &= q^{b'} \pi_{b'} \quad , \quad q^{b'} = Q^{b'} + ix^{b'b} Q_b \\ \rho_x R_\gamma Z^\gamma &= r^{c'} \pi_{c'} \quad , \quad r^{c'} = R^{c'} + ix^{c'c} R_c \end{aligned}\right\} \quad , \tag{4.3}$$

where $p^{a'}$, $q^{b'}$, and $r^{c'}$ are solutions of equation (2.9). Equations (4.1) and (4.3) together give:

$$\rho_x \psi(Z^\alpha) = [p^{a'} \pi_{a'} \, q^{b'} \pi_{b'} \, r^{c'} \pi_{c'}]^{-1} \, . \tag{4.4}$$

Inserting this expression for $\rho_x\ \psi(Z^\alpha)$ into the contour integral formula

$$\phi(x) = \oint \rho_x\ \psi(Z^\alpha)\Delta\pi \tag{4.5}$$

we obtain, upon performing the integral, the result

$$\phi(x) = \lambda[\varepsilon_{a'b'c'}\ p^{a'}\ q^{b'}\ r^{c'}]^{-1} \tag{4.6}$$

with $\lambda = (2\pi i)^2$, the contour being an $S^1 \times S^1$. It is not difficult to verify that $\phi(x)$, as given in (4.6), does indeed satisfy equation (3.6).

Further examples are readily constructed: for instance, using the relations

$$\begin{aligned} \partial\rho_x\ \psi(Z^\alpha)/\partial p^{a'} &= -\pi_{a'}[p^{a'}\ \pi_{a'}]^{-2}\ [q^{b'}\ \pi_{b'}\ r^{c'}\ \pi_{c'}]^{-1} \\ &= -\pi_{a'}\rho_x[P_\alpha Z^\alpha]^{-2}\ [Q_\beta Z^\beta R_\gamma Z^\gamma]^{-1}\ , \end{aligned} \tag{4.7}$$

which follow at once from equations (4.3) and (4.4), one sees that the twistor function

$$[P_\alpha Z^\alpha]^{-2}\ [Q_\beta Z^\beta R_\gamma Z^\gamma]^{-1} \tag{4.8}$$

generates — according to formula (3.2) — the field

$$\phi_{a'}(x) = \lambda\varepsilon_{a'b'c'}\ q^{b'}\ r^{c'}\ [\varepsilon_{a'b'c'}\ p^{a'}\ q^{b'}\ r^{c'}]^{-2}\ , \tag{4.9}$$

and that this field satisfies equations (1.3) and (1.4).

Generalizations of these examples for other values of r, and for higher value of n — in particular, to cases where n is *odd*, for which (as will be discussed elsewhere) a spacetime interpretation is readily available for the resulting formulae — are not difficult to construct.

Notes

1. See, for example, reference [1], section VI.

2. The Grassmann (or Klein) representation for the system of lines in P^3 as a quadric hypersurface in P^5 has been described [2,p.390] as an "... illustration of the use of geometry... that has proved extraordinarily useful and suggestive, as well as being capable of very wide application". These words, written in 1952, are no less true today, especially when one considers the remarkable position occupied by this representation within the framework of twistor theory.

3. The twistor equation is discussed in considerable detail in reference [1], section V. See also references [3], [4], and [5].

4. It is perhaps worth mentioning, as a simple corollary to *Theorem 1*, that a necessary and sufficient condition for a field $\xi^a(x)$ to satisfy equation (2.1) is that it should satisfy

$$\sigma_c \, \sigma_{[a} \nabla_{b]c'} \, \xi^c = 0$$

for all values of σ_a. Another point worth noting is the following. For n = 1 the twistor equation $\nabla_{A'(A} \xi_{B)} = 0$ is, in fact, a special case of a more general differential equation — of wide interest, especially when it is considered within the context of general relativity — known as the *geodesic shearfree condition:* $\xi^A \xi^B \nabla_{A'A} \xi_B = 0$. However, when n > 1 the connection between the twistor equation and the geodesic shearfree condition is lost, the appropriate generalization of the latter being:

$$(\xi_{[a} \nabla_{b]b'} \, \xi_{[c}) \xi_{d]} = 0 \, ,$$

which does indeed, for n = 1, reduce to the geodesic shearfree condition. Solutions of the equation above can be generated (as will be described elsewhere) through the consideration of the intersections of projective n-planes in P^{2n+1} with certain types of complex analytic varieties in P^{2n+1}.

5. Twistor contour integral formulae — alluded to very briefly in Penrose's epoch making 1967 paper [1,p.347] — are first introduced and discussed in detail in reference [3], a magnificent opus from which spring ultimately many of the ideas which have dominated the outlook of this conference and which are destined by their very nature — rooted as they are in the fertile soil of algebraic geometry and complex analysis — to play an increasingly significant role in theoretical physics for many years to come.

6. For an account of the notion of elementary state, see Penrose and MacCallum [4,p.279]. Elementary states are the prototypes of non-singular, positive frequency, asymptotically well-behaved, normalizable wave functions; and they emerge in a strikingly natural way within the context of twistor theory.

ACKNOWLEDGEMENTS I would like to thank R. Penrose, K.P. Tod, R.S. Ward, N.M.J. Woodhouse and other colleagues at The Mathematical Institute, Oxford, for useful conversations in connection with the material described herein. This work has been supported by the Science Research Council, and by a Junior Research Fellowship at Wolfson College, Oxford.

REFERENCES

[1] R. Penrose, Twistor Algebra, J.Math.Phys. 8, 345 (1967).

[2] J.G. Semple and G.T. Kneebone, Algebraic Projective Geometry, Clarendon Press, Oxford (1952).

[3] R. Penrose, Twistor Quantisation and Curved Space-Time, Int. J. Theor. Phys, 1, 61 (1968).

[4] R. Penrose and M.A.H. MacCallum, Twistor Theory: An Approach to the Quantisation of Fields and Space-Time, Phys. Reports, 6C, 241 (1972).

[5] R. Penrose, Twistor Theory: Its Aims and Achievements, in Quantum Gravity: An Oxford Symposium, C.J.Isham, R. Penrose, and D.W. Sciama editors, Clarendon Press, Oxford (1975).

L.P. Hughston
The Mathematical Institute
Oxford
England

C M Patton

Zero rest mass fields and the Bargmann complex structure

My purpose here is to explain certain aspects of the twistor description of zero-rest-mass fields from the point of view of a different complex structure on twistor space. In particular:

(a) the "-2" in "-n-2" where n = 2s, s the helicity.

(b) normalizable fields in terms of holomorphic functions on all of twistor space.

(c) the relative unnaturality of unprimed spin fields.

(d) the inner product on cohomology.

We will begin by considering an inner product structure on holomorphic and antiholomorphic functions on $\mathbb{C}^2$, when $\mathbb{C}^2$ is equipped with a positive definite hermitian scalar product, denoted $\langle , \rangle$. Consider the space

$$F = \{f \in \mathcal{O}(\mathbb{C}^2) \mid \int_{\mathbb{C}^2} |f|^2 \exp - \langle z,z\rangle \, d\mu < \infty\}$$

where $d\mu$ is the normalized Euclidean measure on $\mathbb{C}^2$, i.e., if $\langle z,z\rangle = z_1\bar{z}_1 + z_2\bar{z}_2$, then $d\mu = [-\frac{1}{4}dz_1\wedge d\bar{z}_1\wedge dz_2\wedge d\bar{z}_2]$. Then $F = \bigoplus_{n=0}^{\infty} F_n$ where F_n = {f homogeneous of degree n}. Similarly, we have $\bar{F}$ and $\bar{F}_n$, where $\mathcal{O}$ is replaced by $\bar{\mathcal{O}}$, the antiholomorphic functions.

If $f, g \in F_n$, then we can use the homogeneity to partially integrate the inner product:

$$(f,g) = \int f\bar{g} \exp - \langle z,z\rangle\, i dz_1 \wedge d\bar{z}_1 \wedge dz_2 \wedge d\bar{z}_2 =$$

$$= \left(\int_{\langle z,z\rangle = 1} f\bar{g}\sigma \right) \int_0^\infty t^{2n+3} \exp - t^2 \, dt$$

$$= N(n) \int_\Gamma f\bar{g}(z_1 dz_2 - z_2 dz_1) \wedge (\bar{z}_1 d\bar{z}_2 - \bar{z}_2 d\bar{z}_2)$$

where Γ is an open 2-disc in the sphere $\langle z,z\rangle = 1$ transverse to the fibers of the Hopf fibration $\mathbb{C}^2 - \{0\} \to P^1$ and mapping isomorphically onto P^1 - {a point}. As it stands, the integrand above is not closed. We can, however, make it closed, and still have it agree with the above on $\langle z,z\rangle = 1$, by multiplying the integrand by $\langle z,z\rangle^{-n-2}$. In fact, if $g \in \overline{\mathcal{F}}_n$ then $b(g) \equiv \langle z,z\rangle^{-n-2} g(\bar{z}_1 d\bar{z}_2 - \bar{z}_2 d\bar{z}_1)$ satisfies

(a) $\bar{\partial} b(g) = 0$

(b) $\bar{z}_1 \dfrac{\partial}{\partial \bar{z}_1} + \bar{z}_2 \dfrac{\partial}{\partial \bar{z}_2} \lrcorner\, b(g) = 0$

(c) $z_1 \dfrac{\partial}{\partial z_1} + z_2 \dfrac{\partial}{\partial z_2} \lrcorner\, \partial b(g) = (-n-2) b(g)$

(d) for any $f \in \overline{\mathcal{F}}_n$ $(g,f) = N(n) \displaystyle\int_\Gamma f(z_1 dz_2 - z_2 dz_1) \wedge b(g)$

(e) $b(g)$ is $\bar{\partial}$ exact if and only if $g \equiv 0$.

In particular $\overline{\mathcal{F}}_n \xrightarrow[\approx]{b} H^1(P^1, \mathcal{O}(-n-2))$.

We note that we can include $\mathcal{F}$ and $\overline{\mathcal{F}}$ into $L^2(\mathbb{C}^2, d\mu)$ by $I(f) = f \exp - \frac{1}{2}\langle z,z\rangle$. Then we have the holomorphic and antiholomorphic projectors:

$$P_+ : L^2 \to I(\mathcal{F}) \qquad P_- : L^2 \to I(\overline{\mathcal{F}})$$

$$P_+ f(z) = k \int \exp[-\tfrac{1}{2}\langle z,z\rangle + \langle z,z'\rangle - \tfrac{1}{2}\langle z',z'\rangle] f(z') d\mu'$$

$$P_-f(z) = k \int \exp[-\frac{1}{2}\langle z,z\rangle + \langle z',z\rangle - \frac{1}{2}\langle z',z'\rangle]f(z')d\mu'$$

Similarly, for any complex vector space V equipped with an hermitian inner product $\langle , \rangle_V$, we have associated holomorphic and antiholomorphic projectors, P_{V+}, P_{V-}.

We can think of twistor space $\mathbb{C}^4 = \mathbb{C}^2 \oplus \mathbb{C}^2$ with coordinates $\pi = \begin{pmatrix}\pi_1 \\ \pi_2\end{pmatrix}$ and $\omega = \begin{pmatrix}\omega_1 \\ \omega_2\end{pmatrix}$ with an indefinite hermitian scalar product:

$$A((\pi,\omega), (\pi',\omega')) = i(\omega \cdot \overline{\pi}' - \pi \cdot \overline{\omega}')$$

and a symplectic form $\Omega(p,p')$ on the underlying $\mathbb{R}^8$:

$$\Omega(p,p') = \mathrm{Im}(i(\omega(p) \cdot \overline{\pi(p')} - \pi(p) \cdot \overline{\omega}(p'))).$$

We can, however, put another complex structure on the underlying $\mathbb{R}^8$. Namely, put $\zeta = \frac{1}{\sqrt{2}}(\overline{\pi} + i\overline{\omega})$ and $\nu = \frac{1}{\sqrt{2}}(\pi + i\omega)$ as new complex coordinates. We then have the positive definite hermitian (w.r.t. the new coord.) inner product $B((\zeta,\nu), (\zeta',\nu')) = \zeta \cdot \overline{\zeta}' + \nu \cdot \overline{\nu}'$ with the property that $\mathrm{Im}(\zeta(p) \cdot \overline{\zeta}(p') + \nu(p) \cdot \overline{\nu}(p')) = \Omega(p,p')$, the same symplectic form. Let's call the complex structure with respect to the π,ω coordinates J_0, and with respect to the ζ,ν coordinates, J. Then we can form, given the positive definite inner product, the hilbert space $F(\mathbb{R}^8,J,B)$ of holomorphic functions.

There is a (projective) unitary representation of the group of all real linear transformations of $\mathbb{R}^8$ which leave Ω invariant in this space F which, when restricted to those which also preserve J, is simply the action of coordinate transformations on elements of F. This representation is called the Bargmann-Segal-Fock representation [1]. When this representation is restricted to the (covering of) the conformal group, the space F breaks up into a direct sum of all spin, mass-zero representations, each occurring but once [2].

Comparing the complex structures J and J_0 we find that on $W_+ = \{\zeta=0\}$, $J = J_0$, while on $W_- = \{\nu=0\}$ we have $J = -J_0$. Therefore, an element of $\mathcal{F}$ when restricted to W_+ will be holomorphic w.r.t. J_0 and when restricted to W_- will be antiholomorphic w.r.t. J_0. If we include $\mathcal{F}$ in L^2, we can think of an element as giving us a holomorphic (resp. antiholomorphic) function on every complex (w.r.t. J_0), 2-dimensional subspace, V, of twistor space on which A is positive (resp. negative) definite, by using the holomorphic (resp. antiholomorphic) projector associated with the hermitian (resp. negative of the hermitian) form on V.

$$\delta_V(f) = I^{-1}P_{V^\pm}(I(f)|_V) \quad \text{for} \quad V \text{ in } T^\pm$$

We note that U(2,2), preserving both J_0 and A, has a natural action on the collection of functions $\{\delta_V(f)\}$. Namely, for $g \in U(2,2)$, define $g \cdot \delta_V(f) = (\delta_{g^{-1}V}(f)) \circ g^{-1}$. While this action does not commute with the B-S-F action of g on $\mathcal{F}$, it commutes up to multiplier:

Theorem 1 $g \cdot \delta_V(f) = Q(g,V)\delta_V(g\cdot f)$.

The reason for this multiplier is that we should be dealing with forms rather than functions.

Theorem 2 Let f be any element of $\mathcal{F}(\mathbb{R}^8, B, J)$. Define the infinite collection of functions

$$\{\phi_{B'C'\ldots L'}, \text{ any number of indices, each index taking the values 1 and 2}\}$$

where $\phi_{B'C'\ldots L'}(X) = (K(X)\delta_{V(X)}(f), \bar{\pi}_{B'}\bar{\pi}_{C'}\ldots\bar{\pi}_{L'}\big|_{V(X)})$ for X such that $i(X-X^*)$ is negative definite and $K(X) = \det^{-\frac{1}{2}}(i(X-X^*))\det(I+i\bar{X})$.† Then $\phi_{B'C'\ldots L'}$ satisfy

(a) $\dfrac{\partial}{\partial X^{AA'}}\phi_{B'C'\ldots L'} - \dfrac{\partial}{\partial X^{AB'}}\phi_{A'C'\ldots L'} = 0$

(b) $\dfrac{\partial}{\partial \bar{X}^{AA'}}\phi_{B'C'\ldots L'} = 0.$

Also define $\psi_{BC\ldots L}(X) = (K(X)\delta_{V(X^*)}(f), \pi_B\ldots\pi_L|_{V(X)})$ for the same X. Then the ψ satisfy

(c) $\dfrac{\partial}{\partial X^{AA'}}\psi_{BC\ldots L} - \dfrac{\partial}{\partial X^{BA'}}\psi_{AC\ldots L} = 0$

(d) $\dfrac{\partial}{\partial \bar{X}^{AA'}}\psi_{BC\ldots L} = 0.$

<u>Theorem 3</u> All of the $\phi_{A'B'\ldots L'}$ and $\psi_{AB\ldots L}$ are identically zero if, and only if, f is identically zero.

When f is a monomial in ζ and ν, it is quite straight forward to find the cohomology class which represents the same field information. We will use multi-index notation: $\zeta^p\nu^r = \zeta_p^{p_1}\zeta_2^{p_2}\nu_1^{r_1}\nu_2^{r_2}$, $|r| = r_1 + r_2$, $p! = p_1!p_2!$, etc.

<u>Proposition</u> If $f = a\zeta^p\nu^r$ with $|p| - |r| = n > 0$, and $a \in \mathbb{C}$, then $\psi_{A\ldots L} \equiv 0$ for all choices of indices; $\phi_{A'\ldots L'} \equiv 0$ if the number of indices is not equal to n, and

$$\underbrace{\phi_{A'\ldots L'}}_{n}(X) = \int_{\Gamma\subset V(X)} \pi_{A'}\ldots\pi_{L'}(\pi_{1'}d\pi_{2'} - \pi_{2'}d\pi_{1'}) \wedge \alpha(f)$$

†Here V(X) is the 2-plane associated to the point $x \in \mathbb{CM}$ by the Penrose transform.

Here, $\alpha(f) = N(|p|)a\zeta^p\nu^r(\zeta \cdot \bar{\zeta})^{-|p|-2}(\zeta_1 d\zeta_2 - \zeta_2 d\zeta_1)$, which is a (0,1)-form with respect to the J_0 complex structure and satisfies, again w.r.t. the J_0 complex structure,

(a) $\bar{\partial}\alpha(f) = 0$

(b) $\bar{\gamma} \lrcorner \alpha(f) = 0$

(c) $\gamma \lrcorner \partial\alpha = (-n-2)\alpha(f)$

$$(\gamma = \pi_1 \frac{\partial}{\partial \pi_1} + \pi_2 \frac{\partial}{\partial \pi_2} + \omega_1 \frac{\partial}{\partial \omega_1} + \omega_2 \frac{\partial}{\partial \omega_2}).$$

Therefore $\alpha(f)$, defined on all of T^-, represents a cohomology class $[\alpha(f)] \in H^1(PT^-, \mathcal{O}_{J_0}(-n-2))$. Moreover $[\alpha(f)]$ is the unique class which represents the same field information as f, by virtue of a

Proposition A $\bar{\partial}$ closed (0,1)-form on T^- satisfying (b) and (c) above which is $\bar{\partial}$-exact on every complex 2-dimensional subspace in T^-, is actually exact.

Reversing the roles of ζ and ν we have

Proposition If $f = a\zeta^p\nu^r$ with $|r| - |p| = n > 0$, then

$\phi_{A'\ldots L'} \equiv 0$ for all choices of indices.

$\psi_{A\ldots L} \equiv 0$ if the number of indices is not equal to n.

$$\psi_{A\ldots L}(X) = \int_{\Gamma\subset V(X^*)} \bar{\pi}_A \ldots \bar{\pi}_L \beta(f) \wedge (\bar{\pi}_1 d\bar{\pi}_2 - \bar{\pi}_2 d\bar{\pi}_1).$$

The form $\beta(f) = N(|r|)a\zeta^p\nu^r(\nu \cdot \bar{\nu})^{-|r|-2}(\nu_1 d\nu_2 - \nu_2 d\nu_1)$ is a (1,0) form on T^+ satisfying (w.r.t. J_0)

(a) $\partial\beta(f) = 0$

(b) $\gamma \lrcorner \beta(f) = 0$

(c) $\bar{\gamma} \lrcorner \bar{\partial}\beta(f) = (-n-2)\beta(f)$

and thus represents an element of $H^1(PT^+, \overline{\mathcal{O}}_{J_0}(-n-2))$.

If we want to think of the unprimed information as living on PT^-, we can resort to defining the form

$$\tilde{\beta}(f) = N(|p|) a\zeta^p \nu^r (\zeta \cdot \overline{\zeta})^{-|p|-2} (\zeta_1 d\zeta_2 - \zeta_2 d\zeta_1)$$

which lives on PT^- and satisfies

(a) $\overline{\partial}\tilde{\beta}(f) = 0$

(b) $\overline{\gamma} \lrcorner \tilde{\beta}(f) = 0$

(c) $\gamma \lrcorner \partial\tilde{\beta}(f) = (n-2)\tilde{\beta}(f)$.

We can define fields

$$\tilde{\psi}_{A\ldots L}(X) = \int_{\Gamma \subset V(X)} (\partial\omega_A \ldots \partial\omega_L \tilde{\beta}(f)) \wedge (\pi_1 d\pi_2 - \pi_2 d\pi_1)$$

which again satisfy the field equations, but are not, obviously at least, the same as $\psi_{A\ldots L}$. Here, $\partial\omega_L \tilde{\beta}(f) \equiv \frac{\partial}{\partial\omega_L} \lrcorner \partial\tilde{\beta}(f)$, etc.

Dealing with the primed spin fields first, we can define a pairing on $H^1(PT^-, \mathcal{O}_{J_0}(-n-2))$. If α, α' are (0,1)-forms representing classes $[\alpha]$, $[\alpha']$, let

$$<[\alpha], [\alpha']>_{\mathcal{O}} \equiv \sum_{r \geq 0} \frac{1}{N(|r|+n)r!} \int_{\Gamma \subset V(iI)} (\partial\nu_1^{r_1} \partial\nu_2^{r_2} \alpha) \wedge \overline{(\partial\nu_1^{r_1} \partial\nu_2^{r_2} \alpha')} (\zeta \cdot \overline{\zeta})^{n+|r|+2}$$

where again $(\partial\nu_1)\alpha = \frac{\partial}{\partial\nu_1} \lrcorner \partial\alpha$, etc. This is, in fact, independent of the choice of representatives, α, α'. Moreover, if $f = a\zeta^p \nu^r$, $f' = a'\zeta^s \nu^t$ with $|p| - |r| = n = |s| - |t| > 0$, then

$$(f, f')_F = <[\alpha(f)], [\alpha(f')]>_{\mathcal{O}}.$$

Although I have no proof of it, it is reasonable to believe that for every $f \in F(\mathbb{R}^8, B, J)$, there exists a cohomology class $\tilde{\alpha}_n(f)$ representing the same field information as the $\phi_{\underbrace{A' \ldots L'}_{n}}$ corresponding to f.

Given this, then the closure, H_n, under $\langle , \rangle_O$ of the subspace of classes generated by the $\alpha(f)$, f a monomial of the above type, will actually be contained in $H^1(PT^-, \mathcal{O}(-n-2))$ since the closure, G_n, under $(,)_F$, of the space generated by the f's is actually contained in F. Then $\tilde{\alpha}_n : G_n \to H_n$ coincides on finite sums of monomials with $[\alpha(f)]$ and intertwines the action of U(2,2) on F and $H^1(PT^-, \mathcal{O}(-n-2))$. The space H_n is then the space of normalizable fields.

As a final remark, it may be noted that all of the above may be quite easily generalized to other dimensions. In particular, if we begin with a twistor space $\mathbb{C}^k \oplus \mathbb{C}^k$, we have U(k,k) acting, the analogous hermitian form A, new coordinates ζ, μ, etc. The relevant cohomology in this case is $H^{k-1}(\tilde{PT}^-, \mathcal{O}(-n-k))$. These cohomology classes represent n-fold symmetric fields with k values for each index. The fields will satisfy formally identical field equations on the domain $\{X \mid i(X-X^*)$ neg. definite $k \times k$ matrix$\}$. These, and especially the corresponding inner product on cohomology may prove useful in the multiparticle context.

For further details on all the above, and proofs for the stated propositions and theorems, please see [2], [3], [4] and the references therein.

References

1 V. Bargmann, Group representations on Hilbert spaces of analytic functions, in Analytic Methods in Mathematical Physics, Ed. R. P. Gilbert and R. G. Newton, Gordon and Breach (1968).

2 H. P. Jakobsen and M. Vergne, Wave and Dirac Operators, and Representations of the Conformal Group, J. Functional Analysis, 24, 1, (1977).

3 C. M. Patton, The Metaplectic group and holomorphic projectors, to appear.

4 N. M. J. Woodhouse, Twistor cohomology without sheaves, Twistor Newsletter #2.

Acknowledgment

This work was supported in part by NSF grant MCS77-18723 and an American Mathematical Society Research Fellowship.

C M PATTON
Department of Mathematics
University of Utah
Salt Lake City, Utah 84113

D Burns

Some background and examples in deformation theory

This report is based on lectures given at the Lawrence Conference. Their purpose was to provide some elementary background in deformation theory of complex manifolds and related structures which might be of use in conjunction with the other lectures. I have tried to sketch the (standard) theory behind Penrose's non-linear graviton construction, to give some direction to a general survey. Formally identical techniques are used in Yang-Mills theory, of course, and for a more detailed account in that vein, one should check Rawnsley's lectures [8], or the paper [1] of Atiyah-Hitchin-Singer.

There are several good references now for the basics of complex manifolds, and their deformation theory, e.g., [4], [10]. The reader may also find references there to the original papers of Kodaira and Spencer on the subject. Hopefully, these lectures will offer a quick illustration of the basic techniques of the theory, plausible indications of why the theorems should be true, and some examples and methods of computation.

The author would like to take this opportunity to thank D. Lerner and P. Sommers for their hospitality during the conference, and editorial patience after the conference. Thanks are also due them and many of the participants, who shared their suggestions, opinions and speculations during the conference.

§1. DEFORMATION OF COMPACT COMPLEX MANIFOLDS

To fix ideas, we'll recall that a complex manifold M is a differentiable manifold of real dimension 2n with local complex coordinates $z_\alpha = (z_\alpha^1, \ldots, z_\alpha^n)$ defined on open sets U_α,

$$z_\alpha : U_\alpha \to \mathbb{C}^n$$

related by complex analytic transition functions $f_{\alpha\beta}(z_\beta) = z_\alpha$, $f_{\alpha\beta}$ defined on $z_\beta(U_\alpha \cap U_\beta) \subset \mathbb{C}^n$. The $f_{\alpha\beta}$ satisfy the compatibility condition

$$f_{\alpha\gamma} = f_{\alpha\beta} \circ f_{\beta\gamma},$$

i.e., $f_{\alpha\gamma}(z_\gamma) = f_{\alpha\beta}(f_{\beta\gamma}(z_\gamma))$ on $U_\alpha \cap U_\beta \cap U_\gamma$. A deformation of the complex structure on M is given by making z_α, $f_{\alpha\beta}$ above depend on additional parameters $t = (t^1,\ldots,t^k)$, so we should have

$$z_\alpha(t) = f_{\alpha\beta}(t, z_\beta(t)), \tag{1.1}$$

and

$$f_{\alpha\gamma}(t, z_\gamma(t)) = f_{\alpha\beta}(t, f_{\beta\gamma}(t, z_\gamma(t))), \tag{1.2}$$

and where $z_\alpha(0) = z_\alpha$ is a coordinate for our original complex structure on M. Two complex manifolds M, M' are equivalent ($M \simeq M'$) if there is a diffeomorphism $f: M \to M'$ which is complex analytic.

In practice, one rarely deforms a complex manifold in this fashion.

Example 1: Let $L(\tau)$ be the lattice in $\mathbb{C}$ of complex numbers of the form $m + n\tau$, where m, n are integers, and $\operatorname{Im} \tau > 0$. Identifying z with any point $z + m + n\tau$ gives a torus $M(\tau) = \mathbb{C}/L(\tau)$. It is easy to show that $M(\tau) \simeq M(\tau')$ if and only if $\tau' = \frac{a\tau + b}{c\tau + d}$, with a,b,c,d integers and $ad - bc = 1$.

Example 2: Let $P(x)$ be a homogeneous polynomial of degree d in variables $x = (x_0,\ldots,x_m)$ with complex coefficients, and let $V_P = \{\text{zeroes of } P \text{ in } \mathbb{CP}^n\}$. For the generic P, V_P is a complex manifold, and if we vary the coefficients of P, we get (generally) inequivalent complex manifolds.

Example 3: For $0 < t < 1$, let $A_t = \{z \in \mathbb{C} \mid t < |z| < 1/t\}$. For $t \neq t'$, $A_t \not\simeq A_{t'}$.

We will see more examples of these kinds below, in §'s 1, 2, 3, respectively.

As a first step toward measuring a deformation (1.1), we linearize the compatibility condition (1.2) by differentiating with respect to t at t = 0:

$$\frac{\partial f_{\alpha\gamma}}{\partial t} = \frac{\partial f_{\alpha\beta}}{\partial t} + \frac{\partial f_{\alpha\beta}}{\partial z^i_\beta}\frac{\partial f^i_{\beta\gamma}}{\partial t} . \tag{1.3}$$

Here we set $Z_{\alpha\gamma} = \left.\frac{\partial f_{\alpha\beta}}{\partial t}\right|_{t=0}$ and interpret it as a holomorphic vector field on $U_\alpha \cap U_\gamma$, and similarly for the other terms in (1.3). Rewrite (1.3) as

$$Z_{\alpha\gamma} = Z_{\alpha\beta} + Z_{\beta\gamma} . \tag{1.3)'$$

This is the linearized <u>cocycle condition</u>. If Θ denotes the sheaf of germs of holomorphic vector fields on M, i.e., the infinitesimal complex automorphisms of M, then the vector fields $Z_{\alpha\beta}$ defined on $U_\alpha \cap U_\beta$ determine a Čech 1-cocycle on M with values in Θ.

The deformation (1.1) will be trivial if, for every t,

$$f_{\alpha\beta}(t, h_\beta(t, z_\beta(t))) = h_\alpha(t, z_\alpha(t)), \tag{1.4}$$

for then the maps* $h_\alpha(t, z_\alpha(t))$ of the U_α to themselves patch together to give an equivalence for each t of M_t with M_0, where M_t is the complex manifold with coordinates $z_\alpha(t)$. As before, differentiate (1.4) at t = 0:

$$\frac{\partial f_{\alpha\beta}}{\partial t} + \frac{\partial f_{\alpha\beta}}{\partial z^i_\beta}\frac{\partial h^i_\beta}{\partial t} = \frac{\partial h_\alpha}{\partial t} \tag{1.4)'$$

or, $$Z_{\alpha\beta} + Z_\beta = Z_\alpha \tag{1.5}$$

---

*h_α is holomorphic, for fixed t, with respect to $z_\alpha(t)$.

where Z_α, Z_β are the vector fields $\frac{\partial h_\alpha}{\partial t}$, $\frac{\partial h_\beta}{\partial t}$ defined on U_α, respectively U_β. (1.5) is the coboundary condition.

Recall that the first cohomology group $H^1(M,\Theta)$ of M with coefficients in Θ is given by the quotient of the space of all 1-cocycles, i.e., $\tilde{Z}_{\alpha\beta}$ on $U_\alpha \cap U_\beta$ such that (1.3)' holds, modulo the coboundaries as in (1.5). (For this definition to be precisely correct, it suffices to assume that the U_α are balls in local coordinates--the U_α must be holomorphically convex.)

If M is compact, it is a basic fact that $H^i(M,\Theta)$ is a finite dimensional vector space. The computations above interpret $H^1(M,\Theta)$ as first order infinitesimal deformations of M modulo trivial deformations. The assignment

$$\{f_{\alpha\beta}\} \to \{Z_{\alpha\beta}\} \in H^1(M,\Theta) \tag{1.6}$$

is the derivative of the deformation at $t = 0$. We expect there to be at most dim $H^1(M,\Theta)$ parameters of inequivalent complex structures on M, close to our original one and, in favorable circumstances, this is so. Before discussing why this is so, (and in what sense it is so!), let's analyze another example of deformation, useful in Yang-Mills theory.

Let E be a holomorphic vector bundle over M. Thus we're given invertible, holomorphic $m \times m$ matrix valued functions $g_{\alpha\beta}$ defined on $U_\alpha \cap U_\beta$ which satisfy

$$g_{\alpha\gamma} = g_{\alpha\beta} \cdot g_{\beta\gamma} \quad \text{(pointwise multiplication)} \tag{1.7}$$

on $U_\alpha \cap U_\beta \cap U_\gamma$. To deform E, we allow $g_{\alpha\beta}$ to depend on a parameter t again, with $g_{\alpha\beta}(0) =$ given $g_{\alpha\beta}$. Taking t-derivatives of (1.7) at $t = 0$ gives a 1-cocycle condition on $G_{\alpha\beta} = \frac{\partial g_{\alpha\beta}}{\partial t}$. The derivative of the deformation is the class of $\{G_{\alpha\beta}\}$ in $H^1(M, \text{End}(E))$, where End(E) is the sheaf of (germs of) infinitesimal automorphisms of E, i.e., linear mappings from E to itself.

We'll compute examples of these cohomology spaces below. Two other cohomology spaces play an important role in the theory:

(a) $H^0(M,\Theta)$ = space of 0-cocycles, i.e., collections of vector fields Z_α defined on U_α, so that $Z_\alpha - Z_\beta = 0$ on $U_\alpha \cap U_\beta$. Thus, $H^0(M,\Theta)$ is the space of globally well-defined holomorphic vector fields, the Lie algebra of the group of global complex automorphisms of M.

(b) $H^2(M,\Theta)$ = vector fields $Z_{\alpha\beta\gamma}$ defined on $U_\alpha \cap U_\beta \cap U_\gamma$ with

$$Z_{\alpha\beta\gamma} - Z_{\beta\gamma\delta} + Z_{\gamma\delta\alpha} - Z_{\delta\alpha\beta} = 0$$

on $U_\alpha \cap U_\beta \cap U_\gamma \cap U_\delta$, modulo coboundaries, i.e., modulo $Z_{\alpha\beta\gamma}$ such that

$$Z_{\alpha\beta\gamma} = Z_{\alpha\beta} - Z_{\alpha\gamma} + Z_{\beta\gamma}$$

on $U_\alpha \cap U_\beta \cap U_\gamma$, for some $Z_{\alpha\beta}$'s on $U_\alpha \cap U_\beta$. This space plays the role of obstruction space to the construction of deformations: cf. below. The "favorable circumstances" mentioned above are when $H^2(M,\Theta) = 0$, as stated in the following existence theorem of Kodaira-Spencer-Nirenberg.

Theorem 1 (a) If $H^1(M,\Theta) = 0$, any small deformation of M is trivial.

(b) If $H^1(M,\Theta) \neq 0$, $H^2(M,\Theta) = 0$, then there exists a complex manifold $\mathcal{M}$, parametrizing a family of complex structures on M, of complex dimension = $\dim_{\mathbb{C}} H^1(M,\Theta)$, and such that the derivative mapping described above is an isomorphism of the tangent space of $\mathcal{M}$ at 0 to $H^1(M,\Theta)$.

Remarks: One knows more; namely, in case (b), any small deformation of M can be induced from the one in the theorem by relabelling the parameters.

Much more general results are known about the existence of such "local moduli spaces" $\mathcal{M}$. Theorem 1 is the simplest of its kind, but, to this date, at least, simplifying assumptions analogous to $H^2(M,\Theta) = 0$ have thus far been valid in the cases of physical interest.

In case (a) of Theorem 1, M is called <u>rigid</u>.

To sketch a proof of Theorem 1, let us look at another way to describe the complex structure on a manifold M, and its cohomology. This method examines the Cauchy-Riemann equations on M directly.

If $z^j = x^j + iy^j$ are local holomorphic coordinates on M, set, as usual,

$$dz^j = dx^j + idy^j$$

$$d\bar{z}^j = dx^j - idy^j$$

$$\frac{\partial}{\partial z^j} = \frac{1}{2}\left(\frac{\partial}{\partial x^j} - i\frac{\partial}{\partial y^j}\right)$$

$$\frac{\partial}{\partial \bar{z}^j} = \frac{1}{2}\left(\frac{\partial}{\partial x^j} + i\frac{\partial}{\partial y^j}\right)$$

and $\bar{\partial} f = \dfrac{\partial f}{\partial \bar{z}^1} d\bar{z}^1 + \ldots + \dfrac{\partial f}{\partial \bar{z}^n} d\bar{z}^n.$

$\bar{\partial} f$ is the "anti-holomorphic part" of the total differential df, and $\bar{\partial} f = 0$ if and only if f satisfies the Cauchy-Riemann equations $\dfrac{\partial f}{\partial \bar{z}^j} = 0$, $j = 1,\ldots,n$. Let $\Lambda^{0,k}$ denote the space of <u>kth</u> degree exterior differentials α which can be written

$$\alpha = \Sigma f_{i_1\ldots i_k} d\bar{z}^{i_1} \wedge \ldots \wedge d\bar{z}^{i_k}$$

and define $\bar{\partial}\alpha \in \Lambda^{0,k+1}$ by

$$\bar{\partial}\alpha = \Sigma \frac{\partial f_{i_1\ldots i_k}}{\partial \bar{z}^j} d\bar{z}^j \wedge d\bar{z}^{i_1} \wedge \ldots \wedge d\bar{z}^{i_k}.$$

One has $\bar{\partial}(\bar{\partial}\alpha) = 0$, and <u>locally</u> this is the exact compatibility condition for solving $\bar{\partial}\alpha = \beta$. More generally, if E is a holomorphic vector bundle on M, let $E \otimes \Lambda^{0,k}$ denote the space of (0,k) forms with values in E, so $\alpha \in E \otimes \Lambda^{0,k}$ if α looks locally like

$$\alpha = \Sigma f^s_{i_1\ldots i_k} e_s \otimes d\bar{z}^{i_1} \wedge \ldots \wedge d\bar{z}^{i_k}$$

where $\{e_1,\ldots,e_m\}$ is a local holomorphic basis, or frame, for E. Define

$$\bar{\partial}\alpha = \Sigma \frac{\partial f^s_{i_1\ldots i_k}}{\partial \bar{z}^j} e_s \otimes d\bar{z}^j \wedge d\bar{z}^{i_1} \wedge \ldots \wedge d\bar{z}^{i_k}.$$

Define the kth cohomology group of $\mathcal{O}(E)$, the sheaf of germs of holomorphic sections of E, as

$$H^k(M,\mathcal{O}(E)) = \{\text{globally defined } \beta \in E \otimes \Lambda^{0,k} | \bar{\partial}\beta = 0\}$$

$$\text{modulo}\{\beta = \bar{\partial}\alpha | \alpha \text{ (globally defined)} \in E \otimes \Lambda^{0,k-1}\}.$$

This agrees with our previous definition, using the Čech method. Basically this is because if $\bar{\partial}\beta = 0$, one can solve $\bar{\partial}\alpha = \beta$ locally, on neighborhoods U_α, but these local solutions don't patch correctly on overlaps $U_\alpha \cap U_\beta$. Carrying out a procedure of successive corrections of solutions on multiple overlaps gives the desired isomorphism.

The $\bar{\partial}$-operator from $\Lambda^{0,0}$ = functions to $\Lambda^{0,1}$ "gives" the complex structure on M, since solutions of $\bar{\partial}f = 0$ determine the complex coordinates on M. We can deform M by perturbing the $\bar{\partial}$-operator. To do this, let $\omega \in T \otimes \Lambda^{0,1}$ where T is the (holomorphic) bundle of tangent vectors involving only the $\frac{\partial}{\partial z^j}$, so locally

$$\omega = a^i_j \frac{\partial}{\partial z^i} \otimes d\bar{z}^j.$$

If ω is small, define new Cauchy-Riemann equations on M by

$$\frac{\partial}{\partial \bar{z}^j} f + a^i_j \frac{\partial}{\partial z^i} f = 0, \qquad j = 1,\ldots,n. \tag{1.8}$$

The compatibility condition for finding n independent solutions of these equations is

$$\bar{\partial}\omega + [\omega,\omega] = 0 \tag{1.9}$$

where locally

$$[\omega,\omega] = \{a_k^\ell \frac{\partial a_j^s}{\partial z^\ell} - a_j^\ell \frac{a_k^s}{\partial z^\ell}\}\frac{\partial}{\partial z^s} \otimes d\bar{z}^k \wedge d\bar{z}^\ell \in T \otimes \Lambda^{0,2}.$$

The Newlander-Nirenberg theorem says that this integrability assumption implies the existence of complex coordinate functions which satisfy the new Cauchy-Riemann equations (1.8).

Now think of deforming M by introducing parameters in ω, and write formally

$$\omega = t\omega_1 + t^2\omega_2 + t^3\omega_3 + \ldots$$

Substituting in (1.9) gives successive linear equations

$$\begin{aligned} \bar{\partial}\omega_1 &= 0 \\ \bar{\partial}\omega_2 &= -[\omega_1,\omega_1] \\ &\ldots \\ \bar{\partial}\omega_n &= P_n(\omega_1,\ldots,\omega_{n-1}) \\ &\ldots \end{aligned} \tag{1.10}$$

for ω_n in terms of $\omega_1,\ldots,\omega_{n-1}$. One checks that $\bar{\partial}P_n(\omega_1,\ldots,\omega_{n-1}) = 0$, hence each $P_n(\omega_1,\ldots,\omega_{n-1})$ successively defines a class in $H^2(M,\Theta)$. We see here the role of $H^2(M,\Theta)$ as "obstruction" space: we can solve the nth equation for ω_n if and only if the class of $P_n(\omega_1,\ldots,\omega_{n-1})$ is 0 in $H^2(M,\Theta)$. Conversely, if $H^2(M,\Theta) = 0$, pick any ω_1 with $\bar{\partial}\omega_1 = 0$, and then solve successively for ω_n's to get a <u>formal</u> 1-parameter deformation of M. Convergence is more delicate, of course, and requires compactness of M.

To sketch why a convergent solution exists, use a metric on M to construct Hilbert space inner products on $T \otimes \Lambda^{0,1}$ and let $R(\bar{\partial})^\perp$ be the orthogonal complement of the range of $\bar{\partial}$ from $T \otimes \Lambda^{0,0}$ to $T \otimes \Lambda^{0,1}$. Set $\mathcal{D}(\omega) = \bar{\partial}\omega + [\omega,\omega]$, $\omega \in T \otimes \Lambda^{0,1}$, and note $\bar{\partial}\mathcal{D}(\omega) = 0$, so $\mathcal{D}$ defines a function from $R(\bar{\partial})^\perp$ to $\text{Ker}(\bar{\partial})$ = null space of $\bar{\partial}: T \otimes \Lambda^{0,2} \to T \otimes \Lambda^{0,3}$. If $H^2(M,\Theta) = 0$, $\text{Ker}(\bar{\partial})$ is the range of $\bar{\partial}: T \otimes \Lambda^{0,1} \to T \otimes \Lambda^{0,2}$. The derivative of $\mathcal{D}$ at $\omega = 0$ is just

$\mathcal{D}'(0) = \bar{\partial}$, which maps $R(\bar{\partial})^{\perp}$ onto $\text{Ker}(\bar{\partial})$. By the implicit function theorem, the set of ω near 0 in $R(\bar{\partial})^{\perp}$ with $\mathcal{D}(\omega) = 0$ is a manifold with tangent space at $\omega = 0$ given by the null space of $\mathcal{D}'(0) = \bar{\partial}$. This is identified with $H^1(M,\Theta)$. The compactness of M and the ellipticity of the $\bar{\partial}$-system are used to justify the analysis above.

Remarks: (A) Some modifications are required to drop the assumption $H^2(M,\Theta) = 0$ in the above proof. Then the solutions of $\mathcal{D}(\omega) = 0$ as above don't necessarily form a manifold. This extension is due to Kuranishi. Grauert has extended the theorem to the case where M is allowed to be a singular space.

(B) With minor modifications, the method described here works for deforming complex vector bundles over a fixed M.

Finally, let us compute some easy examples.

(1) The Riemann sphere $= \mathbb{CP}^1$. $H^{\otimes n}$ is the complex line bundle over $\mathbb{CP}^1$ whose holomorphic sections over an open set U are those holomorphic functions f in the homogeneous coordinates x,y for U satisfying $f(\lambda x, \lambda y) = \lambda^n f(x,y)$.

Thus,

$$H^0(\mathbb{CP}^1, \mathcal{O}(H^{\otimes n})) = \begin{cases} 0, \text{ if } n < 0 \\ \text{space of homogeneous polynomials in } x,y \\ \text{of degree } n,\ n \geq 0. \end{cases}$$

To calculate $H^1(\mathbb{CP}^1, \mathcal{O}(H^{\otimes n}))$, we use Serre duality* to conclude

Serre duality says the spaces $H^i(M,\mathcal{O}(E))$ and $H^{n-i}(M,\mathcal{O}(E^*\otimes\Lambda^{n,0}))$ are dual; where M is compact, E is a holomorphic vector bundle, E is the dual bundle and $\Lambda^{n,0}$ is the bundle of complex "volume forms" $a dz^1 \wedge \dots \wedge dz^n$. In terms of forms, if $e \otimes d\bar{z}^{j_1} \wedge \dots \wedge d\bar{z}^{j_i}$ is in $E \otimes \Lambda^{0,i}$, and $e^* \otimes dz^1 \wedge \dots \wedge dz^n \wedge d\bar{z}^{s_1} \wedge \dots \wedge d\bar{z}^{s_{n-i}}$ are annihilated by $\bar{\partial}$, the classes in $H^i(M,\mathcal{O}(E))$ and $H^{n-i}(\mathcal{O}(E^* \otimes \Lambda^{n,0}))$ are paired to the number $\int_M (e,e^*) dz^1 \wedge \dots \wedge dz^2 \wedge d\bar{z}^1 \wedge \dots \wedge d\bar{z}^n$.

$$\dim H^1(\mathbb{CP}^1, \mathcal{O}(H^{\otimes n})) = \dim H^1(\mathbb{CP}^1, (H^{\otimes(-2-n)})).$$

The sheaf Θ is just $\mathcal{O}(H^{\otimes 2})$, so $H^1(\mathbb{CP}^1, \Theta) = 0$ and $\mathbb{CP}^1$ is <u>rigid</u>. ($\mathbb{CP}^1$ doesn't even have discrete changes of complex structure, by the uniformization theorem.) We should also note that $H^1(\mathbb{CP}^1, \mathcal{O}) = 0$, where $\mathcal{O} = \mathcal{O}(H^{\otimes 0})$ is the sheaf of analytic functions on $\mathbb{CP}^1$. For any holomorphic line bundle L on any complex manifold, the infinitesimal deformations of L are given by $H^1(M,\mathcal{O})$, since $\mathcal{O} = \mathcal{O}(\mathrm{End}(L))$--any linear mapping of a line bundle to itself is just multiplication by a holomorphic function. Since $H^1(\mathbb{CP}^1, \mathcal{O}) = 0$, line bundles on $\mathbb{CP}^1$ are rigid. (These are actually only the $H^{\otimes n}$ given above.)

(2) Tori $T = \mathbb{C}/L(\tau)$. For T, the vector field $\frac{\partial}{\partial z}$ is non-vanishing, so $\Theta = \mathcal{O}, \mathcal{O}(\Lambda^{1,0}) = \mathcal{O}$ and $\dim_{\mathbb{C}} H^1(T,\Theta) = \dim_{\mathbb{C}} H^0(T,\mathcal{O}) = 1$, by duality. We have already seen that one parameter, namely τ in example 1 earlier in this §.

(3) $M = \mathbb{CP}^1 \times \mathbb{CP}^1$. Here $\Theta = \Theta_1 \oplus \Theta_2$, where Θ_i is the sheaf of vectors tangent to the $i^{\underline{th}}$ factor, $i = 1, 2$. The <u>Künneth formula</u>* tells us $H^1(M,\Theta) = H^1(M,\Theta_1) + H^1(M,\Theta_2) =$ two copies of $H^1(\mathbb{CP}^1, \Theta) \otimes H^0(\mathbb{CP}^1, \ )$ $+ H^0(\mathbb{CP}^1, \Theta) \otimes H^1(\mathbb{CP}^1, \mathcal{O}) = 0$, by the earlier example.

---

*The <u>Künneth theorem</u> says that, if E_i is a holomorphic vector bundle on M_i, $i = 1, 2$, and E is the bundle $p_1^*E_1 \times p_2^*E_2$ on $M = M_1 \times M_2$ (where $p_i\colon\ M \to M_i$ is the projection onto the $i^{\underline{th}}$ factor), Then $H^i(M,\mathcal{O}(E)) = \bigoplus_{k+\ell} H^k(M_1,\mathcal{O}(E_1)) \otimes H^\ell(M_2,\mathcal{O}(E_2))$. The map is given, in terms of differential forms,

$$(\sigma_1 \otimes \omega_1) \times (\sigma_2 \otimes \omega_2) \to (p_1^*\sigma_1 \otimes p_2^*\sigma_2) \otimes p_1^*\omega_1 \wedge p_2^*\omega_2.$$

Here σ_j is (locally) a differentiable section of E_i, and ω_j are $(0,k)$ or $(0,\ell)$ forms.

§2. RELATIVE DEFORMATION THEORY

Suppose now that M is a compact, complex manifold, and that $M \subset \bar{M}$, where $\bar{M}$ is another complex manifold, not necessarily compact. We have a short exact sequence of holomorphic vector bundles on M:

$$0 \to T_M \to T_{\bar{M}}\big|_M \to N \to 0 \tag{2.1}$$

where T_M = bundle of holomorphic tangents to M, $T_{\bar{M}}\big|_M$ = tangents to $\bar{M}$, tangent at points $m \in M$, and N = normal vectors to M in $\bar{M}$. We get a long exact sequence from (2.1):

$$\cdots \to H^0(\Theta_{\bar{M}}\big|_M) \overset{\alpha}{\to} H^0(\mathcal{O}(N)) \xrightarrow{\delta_1} H^1(\Theta_M) \to H^1(\Theta_{\bar{M}}\big|_M) \to H^1(\mathcal{O}(N)) \xrightarrow{\delta_2} H^2(\Theta_M) \to \cdots$$

Here we've omitted mention of M in $H^i(M,\Theta_M)$, etc., and used Θ_M for $\mathcal{O}(T_M)$, $\Theta_{\bar{M}}\big|_M$ for $\mathcal{O}(T_{\bar{M}}\big|_M)$. At the infinitesimal level, deformations of M <u>inside</u> $\bar{M}$ will be measured by $H^0(\mathcal{O}(N))$, a normal holomorphic vector field along N giving a first order "motion" of M in $\bar{M}$. Some fields come via α from holomorphic fields on $\bar{M}$, so they give trivial deformations of the "abstract" manifold M when considered in $H^1(\Theta_M)$ via δ_1. The obstructions to realizing a given $\sigma \in H^0(\mathcal{O}(N))$ as the deirvative of a deformation in $\bar{M}$ lie in $H^1(\mathcal{O}(N))$, and δ_2 of these obstructions measures whether one may deform the structure on M by moving M "outside" of $\bar{M}$.

Let us first sketch the basic set-up, in a Čech formalism, as follows: We'll be given a covering $\{U_\alpha\}$ of M, and we'll want varying coordinate functions $z_\alpha(t)$, and transition functions $f_{\alpha\beta}(t,z_\beta(t))$ satisfying (1.3), and imbeddings $h_\alpha(t,z_\alpha(t))$ of U_α into $\bar{M}$, holomorphic at "time" t with respect to

$z_\alpha(t)$, with $h_\alpha(0,z_\alpha(0))$ the identity imbedding. These are all to satisfy the condition

$$h_\alpha(t,z_\alpha(t)) = h_\beta(t,f_{\alpha\beta}(t,z_\beta(t))) \tag{2.2}$$

on $U_\alpha \cap U_\beta$. As before, $\frac{\partial h_\alpha}{\partial t}$ should be the derivative of the deformation, but read only modulo tangent vectors to M, since moving M tangent to itself doesn't change the structure of M as complex submanifold of $\bar{M}$. At $t = 0$, we get:

$$\frac{\partial h_\beta}{\partial t} = \frac{\partial h_\alpha}{\partial t} + \frac{\partial h_\alpha}{\partial z_\alpha^k} \cdot \left(\frac{\partial f_{\alpha\beta}^k}{\partial t} + \frac{\partial f_{\alpha\beta}^k}{\partial z_\beta^\ell}\frac{\partial z^\ell}{\partial t}\right) \tag{2.2$'$}$$

$$\equiv \frac{\partial h_\alpha}{\partial t} \quad \text{mod tangents to M,}$$

since $h_\alpha(0,z_\alpha(0))$ is the identity. (2.2)' implies $\frac{\partial h_\alpha}{\partial t}$, read modulo tangents, gives a globally defined normal vector to M in $\bar{M}$, i.e., a section $\sigma \in H^0(\mathcal{O}(N))$. Conversely, given such a σ, it determines h_α's, to first order, in the normal direction. Once we realize (locally) $\frac{\partial h_\alpha}{\partial t} = \tilde{Z}_\alpha$, $\tilde{Z}_\alpha$ a holomorphic section of $\Theta_{\bar{M}}|_M$, we've determined $\frac{\partial f_{\alpha\beta}}{\partial t}$ at $t = 0$ in (2.2). (This is just computing δ_1 of σ.)

The first obstruction occurs when we try to determine the second order expansion of h_α in t. Taking $\frac{\partial^2}{\partial t^2}$ at $t = 0$ of (2.2) gives

$$\frac{\partial^2 h_\alpha}{\partial t^2} - \frac{\partial^2 h_\beta}{\partial t^2} + 2\,\frac{\partial^2 h_\alpha}{\partial z_\alpha^\ell \partial t}\,\frac{\partial f_{\alpha\beta}^\ell}{\partial t} \equiv 0, \tag{2.2$''$}$$

mod tangents to M. Here the expression $\tilde{Z}_{\alpha\beta}^{(2)} = \frac{\partial^2 h_\alpha}{\partial z_\alpha^\ell \partial t} \cdot \frac{\partial f_{\alpha\beta}^\ell}{\partial t}$ on $U_\alpha \cap U_\beta$ is a one cocycle for $\mathcal{O}(N)$, i.e., when read modulo tangents to M. Note that it is determined by our first order choices already made. If $H^1(\mathcal{O}(N)) = 0$, we can write $\tilde{Z}^{(2)}_{\alpha\beta} = -\tilde{Z}^{(2)}_{\alpha} + \tilde{Z}^{(2)}_{\beta}$ mod tangents, and satisfy (2.2)" by choosing

$$\frac{\partial^2 h_\alpha}{\partial t^2} = \tilde{Z}^{(2)}_{\alpha}.$$

The analogue of Theorem 1 is:

Theorem 2 (Kodaira) If $H^1(\mathcal{O}(N)) = 0$, then there exists a d(complex)-parameter family of deformations of M inside $\bar{M}$, where $d = \dim_{\mathbf{C}} H^0(\mathcal{O}(N))$. The derivative mapping is an isomorphism of the tangent space to the parameters at 0 with $H^0(\mathcal{O}(N))$.

Remarks: The non-linear operator proof of this statement is harder to set up than the previous one (Theorem 1). This is because the operator, unlike the $\mathcal{D}$ of §1, must involve to all orders the imbedding of M in $\bar{M}$. We've already seen this in the first Čech obstruction: the second derivatives in $\tilde{Z}^{(2)}_{\alpha\beta}$ depend on the imbedding, although $H^0(\mathcal{O}(N))$ and $H^1(\mathcal{O}(N))$ depend only on the (first order) normal bundle N over M.

The non-linear operator method does work, and has the advantage of applying to other situations, for example the parametrization of nearby minimal surfaces in the calculus of variations, where no cohomological or sheaf-theoretical arguments are available.

Douady has given a much more general form of this theorem.

Of importance to us is the following corollary, which deals with a "doubly-relative" situation: suppose we now continuously deform $\bar{M}$ to $\bar{M}_s$ for some (small) complex parameter s. Do we have a continuous deformation M_s of M, M_s in $\bar{M}_s$, and if M has a d-parameter family of deformations in $\bar{M}$, does M_s have such a family in $\bar{M}_s$? The answer to both questions is yes, if $H^1(M,\mathcal{O}(N)) = 0$. This follows as an immediate corollary of Theorem 2 if $H^1(M,\mathcal{O}) = 0$, e.g., for $M = \mathbf{CP}^1$ or $\mathbf{CP}^1 \times \mathbf{CP}^1$. To see this let D be a small parameter ball in $\mathbb{C}^r$, and let $\mathcal{M} = \bigcup_{s \in D} \bar{M}_s$ be the "total space" of the deformation of $\bar{M}$. We have a short exact sequence of normal bundles on M:

$$0 \to N \to N_{M,\mathcal{M}} \to N_{\bar{M},\mathcal{M}}\big|_M \to 0, \tag{2.3}$$

where $N_{M,\mathcal{M}}$ is the normal bundle of M in $\mathcal{M}$, and $N_{\bar{M},\mathcal{M}}\big|_M$ is the normal bundle to $\bar{M}$ inside $\mathcal{M}$, restricted to M. $N_{\bar{M},\mathcal{M}}\big|_M$ is isomorphic to the tangent bundle T_D of D pulled back to $\mathcal{M}$, so is trivial, and $\mathcal{O}(N_{\bar{M},\mathcal{M}}\big|_M)$ is isomorphic to $\bigoplus_{r\text{ copies}} \mathcal{O}$. In cohomology, then, we get

$$H^1(\mathcal{O}(N)) \to H^1(\mathcal{O}(N_{M,\mathcal{M}})) \to \bigoplus_r H^1(\mathcal{O}).$$

Since the outer terms are assumed 0, so is $H^1(\mathcal{O}(N_{M,\mathcal{M}}))$. Hence, Kodaira's theorem applies to M in $\mathcal{M}$. But $\dim H^0(\mathcal{O}(N_{M,\mathcal{M}})) = \dim H^0(\mathcal{O}(N)) + r$, $r = \dim \bigoplus_r H^0(\mathcal{O})$. Since we already know $d = \dim H^0(\mathcal{O}(N))$ of the parameters give the deformations in $\bar{M}$, we know that the remaining r must give deformations of M into all of the nearby $\bar{M}_s$. Further, each $\bar{M}_s$, for s small, must contain a d-parameter family of deformations of M. (Note that each small deformation M_t of M must lie entirely in one of the $\bar{M}_s$, since M_t is still compact and the usual coordinate functions $z_1,\ldots,z_r$ on D must be constant on M_t, when pulled back to $\mathcal{M}$.)

As examples of the above, let us make some of the basic calculations used in the construction of non-linear gravitons or the "non-self dual twistor transform."

(1) For the non-linear graviton in curved twistor space, we want to let M = a line in $\mathbb{CP}^3$, $\bar{M}$ = an open neighborhood of that line. The normal bundle N will be two-dimensional, and since M is the intersection of two planes E_1 and E_2 in $\mathbb{CP}^3$, N is the sum of two line bundles N_i of normal vectors to M in E_i, i = 1, 2. Each N_i is just the $H = H^{\otimes 1}$ of §1, since

$E_i = \mathbb{CP}^2$, and in homogeneous coordinates $[x,y,z]$, if M is given by $x = 0$, $z\frac{\partial}{\partial y}$ is tangent vector on $\mathbb{CP}^2$ which is not tangential to M along M except where $z = x = 0$, i.e., at one point in M. Since only linear functions on $\mathbb{C}^2$ have one simple zero when thought of as sections on $M = \mathbb{CP}^1$, N_i must be H. Hence,

$$H^0(M,\mathcal{O}(N)) = H^0(M,\mathcal{O}(N_1)) + H_1^0(M,\mathcal{O}(N_2)) = \mathbb{C}^4,$$

$$H^1(M,\mathcal{O}(N)) = 0.$$

The set of sections $\sigma \in H^0(M,\mathcal{O}(N))$ which vanish somewhere on M form the complex light cone in $\mathbb{C}^4$.

If we deform the ambient $\bar{M}$, we first note that $H^1(M,\mathcal{O}) = 0$, so that the corollary to Theorem 2 applies. Since $H^1(M,\Theta) = 0$, all the deformations of M inside $M = \bigcup_s \bar{M}_s$ must be isomorphic to M. Hence, the normal bundles N_t = normal bundle of M_t in $\bar{M}_s$ (s determined by t) give a deformation of the original N on the fixed manifold M. Let us compute the deformations of that N: End(N) is the bundle $(H \oplus H) \otimes (H^{\otimes -1} \oplus H^{\otimes -1})$, so

$$H^i(M,\mathcal{O}(\mathrm{End}(N))) = \bigoplus_{4\ \text{copies}} H^i(M,\mathcal{O}) = \begin{cases} \mathbb{C}^8, & i = 0 \\ 0, & i = 1 \end{cases}.$$

Hence, N is also rigid, and the infinitesimal incidence relation $\{\sigma \in H^0(M_t,\mathcal{O}(N_t)) | \sigma \text{ vanishes somewhere on } M_t\}$ is once again a non-degenerate quadratic cone, which varies holomorphically with t. This gives the holomorphic conformal structure in Penrose's construction.

(2) For the non-self-dual transform, we consider $\mathbb{CP}^3 \times \mathbb{CP}^3$ with homogeneous coordinates x_i, y_i, $i = 1,\ldots,4$, on the respective factors, and $\bar{M} \subset \mathbb{CP}^3 \times \mathbb{CP}^3$ given by $\Sigma x_i y_i = 0$. Let $\mathbb{CP}^1 \times \mathbb{CP}^1 = M \subset Q$,

e.g., $\{x_1 = x_2 = 0\} \times \{y_3 = y_4 = 0\}$. The normal bundle N is harder to compute exactly here, and we'll just compute its cohomology using exact sequences. First, let's denote by $\mathcal{O}(k,\ell)$ the sheaf of sections of the line bundle on M obtained by tensoring $H^{\otimes k}$ on the first factor with $H^{\otimes \ell}$ on the second. We've an exact sequence of normal bundles again:

$$0 \to N \to N_2 \to N_3 \to 0 \tag{2.4}$$

where N_2 = normal bundle of M in $\mathbb{CP}^3 \times \mathbb{CP}^3$,

N_3 = normal bundle of $\bar{M}$ in $\mathbb{CP}^3 \times \mathbb{CP}^3$, restricted to M.

$\mathcal{O}(N_2)$ is just $\mathcal{O}(1,0) + \mathcal{O}(1,0) + \mathcal{O}(0,1) + \mathcal{O}(0,1)$, arguing as before. $\mathcal{O}(N_3)$ is just $\mathcal{O}(1,1)$, basically because the form $\Sigma x_i y_i$ is quadratic, linear in the x's and the y's. Consider the cohomology of (2.4):

$$0 \to H^0(\mathcal{O}(N)) \to H^0(\mathcal{O}(N_2)) \overset{\beta}{\to} H^0(\mathcal{O}(N_3)) \overset{\delta}{\to} H^1(\mathcal{O}(N)) \to H^1(\mathcal{O}(N_2)) \to \ldots \tag{2.5}$$

In the following calculations we use the Künneth formula

$$H^0(\mathcal{O}(N_2)) = \bigoplus_{\text{4-copies}} H^0(\mathbb{CP}^1, \mathcal{O}(H)) \times H^0(\mathbb{CP}^1, \mathcal{O})) = \mathbb{C}^8$$

$$H^1(\mathcal{O}(N_2)) = 0$$

$$H^0(\mathcal{O}(1,1)) = H^0(\mathbb{CP}^1, \mathcal{O}(H)) \otimes H^0(\mathbb{CP}^1, \mathcal{O}(H)) = \mathbb{C}^4$$

To apply Kodaira's theorem, we want $H^1(\mathcal{O}(N)) = 0$, and to recover complex conformal Lorentz manifolds we again want $H^0(\mathcal{O}(N)) = \mathbb{C}^4$ with a holomorphically varying cone of infinitesimal incidence cones which are non-degenerate quadratic cones. Both dimension counts work out if we check β is surjective in (2.5). If $[x,y]$ and $[u,v]$ are homogeneous coordinates on the first and second factors of M, then an element σ of $H^0(\mathcal{O}(N_2))$ is identified with a

4-tuple of linear forms (f_1,f_2,g_1,g_2), the f_i in x and y, the g_i in u and v. An element $\tau \in H^0(\mathcal{O}(N_3))$ is a quadratic form in x,y,u,v, linear in x,y and in u,v. The map β is, up to linear change of the homogeneous variables, given by

$$\beta(f_1,f_2,g_1,g_2) = uf_1 + vf_2 + xg_1 + yg_2.$$

This is obviously surjective.

As noted earlier, M is rigid, and so analogous to Example 1 just above, we now have only to show the normal bundle N is rigid on M. We won't carry that out here, but it can be done by comparison with N_2 and N_3 again.

Remarks: In the full non-linear graviton construction one wants to know not just the conformal structure but rather the actual set of deformed curves or surfaces, and their intersection relations. For this we refer the reader for examples to either the paper of Curtiss-Lerner-Miller [2] or Ward [9]. For more on Example 2, cf. the papers of Isenberg-Yasskin-Green [3] or Witten [11].

§3. CONCLUDING REMARKS

The deformation theory outlined above is sufficient for setting up the local formulation of non-linear gravitons. The global theory, which appears not to exist now, is much more difficult, especially since the problem is not very precisely formulated, so far as I understand the state of the art. Roughly speaking, one wants global deformations of "all" of $\mathbb{P}\mathbb{T}^+$ or $\mathbb{P}\mathbb{T}^-$ to give a truly global non-linear graviton. I say "all" mainly because it is extremely unclear what one wants to do with the boundary N of $\mathbb{P}\mathbb{T}^\pm$.

In general, very little in the way of existence theorems is known for deformations of open complex manifolds M with boundaries ∂M. Various works of Hamilton, Kiremidjian and Stanton appear to have the fault that the boundary conditions imposed appear either too strong for what is needed in twistor theory, or else the assumptions on eigenvalues of the Levi-form of the boundary are not met by N.

One might try sticking to the compact manifold N and deforming it as a CR-manifold, i.e., deforming its partial complex structure it inherits from the ambient $\mathbb{P}\mathbb{T}$. One loses ellipticity, however, in the resulting system--the $\bar{\partial}_b$-system--of equations on N. Further, one does not even know in the most favorable cases (positive definite Levi-form, and dimension ≥ 5) whether the integrability theorem analogous to the Newlander-Nirenberg theorem is true. Further, still, the N of twistor theory isn't even a favorable case!

It seems to me (which is not much of a recommendation for what follows) that the global formulation of the problem required here is very akin to the perhaps simpler problem of superposing "right" and "left" solutions of the linearized field equations, and the "norming" of such fields to give some sort of probability amplitudes. The boundary conditions needed will hopefully be determined by what is wanted physically, with the aid, perhaps, of some insight into particular solutions--much like the origins of twistor theory in the expression of solutions of field equations by contour integrals.

Finally, I apologize for not mentioning deformations of singular varieties enough to help introduce the reader to the "Euclidean gravitons" constructed by Hitchin.

References

1 M. F. Atiyah, N. Hitchin, I. M. Singer, Self-duality in four-dimensional Riemannian geometry, Proc. Roy. Soc. Lond., 1978.

2 W. D. Curtis, D. E. Lerner, F. R. Miller, Complex pp-waves and non-linear gravitons, to appear in GRG.

3 J. Isenberg, P. Yasskin, P. Green, Non-Self-dual gauge fields, Preprint (U. of Maryland).

4 K. Kodaira, Stability of compact complex submanifolds, Am. J. of Math., 1963.

5 K. Kodaira and J. Morrow, Complex Manifolds, Holt-Rinehart, 1972.

6 M. Kuranishi, New proof of the existence of deformations of complex structure, in Complex Analysis, Springer, 1965.

7 R. Penrose, Nonlinear gravitons and curved twistor theory, Gen. Rel. Grav., 1976.

8 J. Rawnsley, Differential geometry of instantons, preprint (Dublin Institute for Advanced Studies).

9 R. S. Ward, A class of self-dual solutions of Einstein's equations, Proc. Roy. Soc. Lond., 1978.

10 R. O. Wells, Differential analysis on complex manifolds, Prentice-Hall, 1972.

11 E. Witten, An interpretation of classical Yang-Mills theory, preprint (Harvard University.)

D BURNS
Department of Mathematics
University of Michigan
Ann Arbor, Michigan 48104

E T Newman

Deformed twistor space and H-space

In this note, we will describe how the space of solutions of a second order differential equation, the so-called good cut equation, defines two interesting manifolds, deformed twistor space [1,2] and H-space [3,4], and further how these solutions induce on these spaces in a natural manner a Kähler and a complex Riemannian metric, respectively.

We begin by considering $S^2 \times S^2$ as a complex manifold coordinatized by the complex stereographic coordinates $(\zeta,\tilde{\zeta})$. (The sphere metric for an S^2 in these coordinates is

$$ds^2 = 4\,\frac{d\zeta d\bar{\zeta}}{(1+\zeta\bar{\zeta})^2} \tag{1}$$

with bar denoting complex conjugate.) Actually we will be concerned not with the entire $S^2 \times S^2$ but with the anti-holomorphic strip defined by an open neighborhood around the line $\tilde{\zeta} = \bar{\zeta}$, i.e. we will be considering the complex thickening of the real sphere so that $\tilde{\zeta}$, though still independent of ζ, has its values close to $\bar{\zeta}$. This region will be referred to as $S^2_{\mathbb{C}}$.

The good cut equation [3] for the holomorphic function $Z(\zeta,\tilde{\zeta})$ (defined on $S^2_{\mathbb{C}}$) has the form

$$\frac{\partial}{\partial\zeta}\,(1+\zeta\tilde{\zeta})^2\frac{\partial}{\partial\zeta}Z = \sigma^\circ(Z,\zeta,\tilde{\zeta}) \tag{2}$$

with σ° an arbitrary spin 2, conformal weight -1 holomorphic function in the three variables $Z,\zeta,\tilde{\zeta}$. (For the meaning of spin and conformal weight, see Appendix A.)

(As an aside, we mention that the good cut equation has its origin in a physical question [3]. In Minkowski space a light cone emanating from a point has a vanishing asymptotic shear or distortion. The question naturally arose, does an arbitrary asymptotically flat physical space-time have any null surfaces which resemble Minkowski space null cones in the sense of having a vanishing asymptotic shear? The answer in general is no--however if the physical space is analytic in the neighborhood of null infinity, by thickening the real space into the complex, one can find complex null surfaces which are asymptotically shear free. In fact it is the good cut equation which determines these surfaces, with every regular solution $Z(\zeta,\tilde{\zeta})$ on $S^2_{\mathbb{C}}$ corresponding to such a surface. The $\sigma^{\circ}(Z,\zeta,\tilde{\zeta})$ in the good cut equation is a measure of the strength of the gravitational radiation associated with the asymptotically flat space-time. For Minkowski space, where there is no radiation, one has $\sigma^{\circ} = 0$.)

There are two points of view we can adopt towards solving the good cut equation. It can

a) be thought of [5] as an ordinary 2nd order differential equation with $\tilde{\zeta}$ as a parameter, with the general solution thus depending on two arbitrary constants (say $\tilde{\alpha}$ and $\tilde{\beta}$) plus $\tilde{\zeta}$, i.e. $u = Z(\zeta;\tilde{\zeta},\tilde{\alpha},\tilde{\beta})$, or

b) the $\tilde{\alpha}$ and $\tilde{\beta}$ can be thought of [3,4] as functions of $\tilde{\zeta}$ which must be chosen subject to the condition that $Z(\zeta,\tilde{\zeta})$ be holomorphic on $S^2_{\mathbb{C}}$.

These points of view can be illustrated with the good-cut equation from Minkowski space, i.e. with $\sigma^{\circ} = 0$. Thus

$$\frac{\partial}{\partial\zeta}(1+\zeta\tilde{\zeta})^2\frac{\partial}{\partial\zeta}Z = 0 \tag{3}$$

and hence

$$Z = \frac{\tilde{\alpha}\zeta+\tilde{\beta}}{1+\zeta\tilde{\zeta}} , \tag{4}$$

which is the general solution from the first point of view. If one now lets

$$\begin{aligned} \tilde{\alpha} &= a\tilde{\zeta}+b \\ \tilde{\beta} &= c\tilde{\zeta}+d \end{aligned} \tag{5}$$

then

$$Z = \frac{a\zeta\zeta+b\zeta+c\tilde{\zeta}+d}{1+\zeta\tilde{\zeta}} , \tag{6}$$

with a,b,c and d being constants.

It is not difficult to convince oneself that the choice (5) yields the most general solution to (3) which is holomorphic on $S^2_{\mathbb{C}}$.

For future reference we point out that (6) may be rewritten in the forms

$$Z = \sum_{\ell,m} x_{\ell m} Y_{\ell m} (\zeta,\tilde{\zeta}) \tag{7}$$

with the sum going over the first four spherical harmonics and

$$Z = x^a \ell_a (\zeta,\tilde{\zeta}) \tag{8}$$

$$\ell_a = 2^{-\frac{1}{2}}(1+\zeta\tilde{\zeta})^{-1}(1+\zeta\tilde{\zeta}, \zeta+\tilde{\zeta}, \frac{\zeta-\tilde{\zeta}}{i} , -1+\zeta\tilde{\zeta}) , \tag{9}$$

where $x_{\ell m} = (x_{00}, x_{1-1}, x_{10}, x_{11})$ and $x^a = (x^0,x^1,x^2,x^3)$ are linear combinations of the a,b,c,d and ℓ_a can be thought of as a complex Minkowski space null vector for each value of ζ and $\tilde{\zeta}$. As ζ and $\tilde{\zeta}$ move over $S^2 \times S^2$, ℓ_a spans the entire complex Minkowski space null cone.

Note that from point of view a) the general solution to (3) depends on three arbitrary complex parameters $\tilde{\alpha}, \tilde{\beta}, \tilde{\zeta}$ and the solution space thus forms a three complex dimensional set while from b) the solution depends on four arbitrary complex parameters x^a and thus forms a four complex dimensional manifold. We will see that this result holds true for the general good cut equation with point of view a) leading to deformed twistor space and b) to H-space.

The two points of view may be visualized as a) a line and b) a surface (ruled by the lines) in a complex three dimensional space. See figure 1.

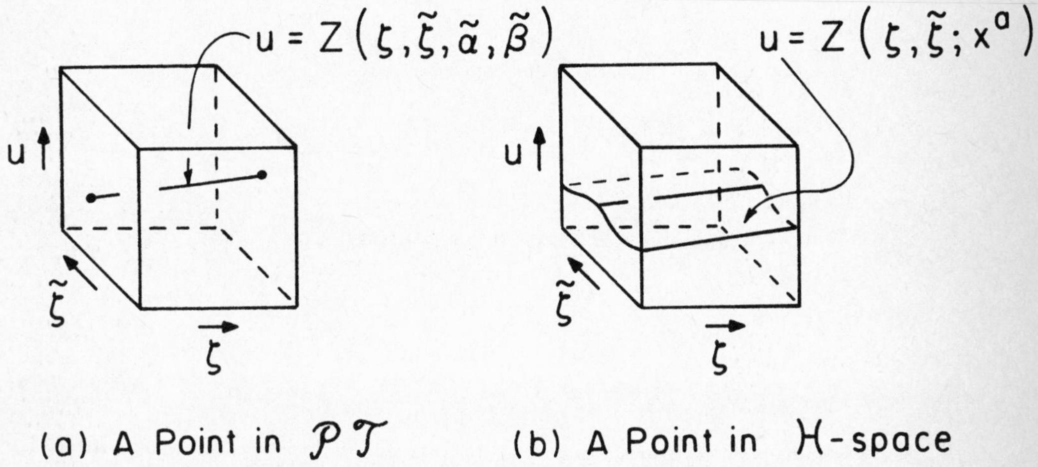

Figure 1

Twistor Space

From the first point of view the general solution of (2) can be written $u = Z(\zeta;\tilde{\zeta},\tilde{\alpha},\tilde{\beta})$ where $\tilde{\alpha}$ and $\tilde{\beta}$ can be, for example, the "initial" position and slope of the curve at $\zeta = \zeta_0$ and $\tilde{\zeta}$ defines the plane in which the curve lies. $\tilde{\alpha}, \tilde{\beta}, \tilde{\zeta}$ are to be thought of as the local coordinates of the three complex dimensional manifold of these curves. Knowledge of the solution $u = Z(\zeta)$ allows one to find the "final" position and slope at $\zeta = \zeta_1$ which could also be used as local coordinates. The solution itself thus acts to define the transition functions for the manifold. This results in this space being identical to the deformed projective twistor space of Penrose, PT.

Deformed twistor space T [1,2,5] can be defined by introducing homogeneous coordinates

$$w^{0'} = i\tilde{r}\tilde{\alpha},\quad w^{1'} = -i\tilde{r}\tilde{\beta}$$

$$w^{2'} = \tilde{r},\quad w^{3'} = -\tilde{r}\tilde{\zeta}$$

and the transition functions induced from PT. Note that $u = Z(\zeta, w^{\alpha'})$ is homogeneous of degree zero in $w^{\alpha'}$.

We consider also the conjugate equation to the good cut equation, namely

$$\frac{\partial}{\partial\tilde{\zeta}}(1+\zeta\tilde{\zeta})^2\frac{\partial}{\partial\tilde{\zeta}}\tilde{Z} = \tilde{\sigma}^o(\tilde{Z},\tilde{\zeta},\zeta)\ , \tag{10}$$

where $\tilde{\sigma}^o(\tilde{Z},\tilde{\zeta},\zeta)$ is obtained by analytically extending the complex conjugate of σ^o away from the surface where $\tilde{\zeta} = \bar{\zeta}$. The solutions

$$u = \tilde{Z}(\tilde{\zeta}, z^{\alpha}) \tag{11}$$

thus define a manifold $\tilde{T}$ conjugate to T. (It should be pointed out that we have taken a slight liberty here with the conventional notation. What we have here called twister space T is conventionally called dual twistor space and $\tilde{T}$ is the conventional twistor space.)

We will now define on the product manifold $\tilde{T} \times T$, in a neighborhood of the diagonal

$$w^{\alpha'} = \overline{z^{\alpha}}$$

(the neighborhood being referred to as CK and the diagonal as K), the function

$$\Sigma(z^{\alpha}, w^{\alpha'}) = ir\tilde{r}(1+\zeta\tilde{\zeta})\{\tilde{Z}(\tilde{\zeta}, z^{\alpha}) - Z(\zeta, w^{\alpha'})\},$$

or

$$\Sigma(z^{\alpha},w^{\alpha'}) = i(z^2w^{2'} + z^3w^{3'}) \{\tilde{Z}(-\frac{w^{3'}}{w^{2'}},z^{\alpha}) - Z(-\frac{z^3}{z^2},w^{\alpha'})\}, \tag{12}$$

which is obviously homogeneous of degree (1,1) in z^{α} and $w^{\alpha'}$. Σ, by being considered as the Kähler scalar and hence defining on K the hermitian metric

$$G_{\alpha\beta'} = \left.\frac{\partial^2\Sigma}{\partial z^{\alpha}\,\partial w^{\beta'}}\right|_{w^{\beta'} = \overline{z^{\beta}}} \tag{13}$$

with

$$ds^2 = G_{\alpha\beta'}\,dz^{\alpha}dw^{\beta'}\Big|_{w^{\beta'} = \overline{z^{\beta}}}$$

makes K into a Kähler manifold. The signature is (++--).

The Kähler scalar, which is not usually a geometric quantity, is so in our case. It measures (with an appropriate scaling) the difference in u values between a curve $u = Z(\zeta,\tilde{\zeta}_0)$ and its conjugate curve $u = \tilde{Z}(\tilde{\zeta},\overline{\tilde{\zeta}_0})$ along the common generator $\zeta = \overline{\tilde{\zeta}_0}$, $\tilde{\zeta} = \tilde{\zeta}_0$.

The Kähler manifold obtained in this manner has many remarkable features, the most surprising being that the Ricci tensor vanishes and that the Weyl tensor has only three independent non-vanishing components--these components being identical to the radiation components of the original physical space Weyl tensor.

In the special case of $\sigma^{\circ} = 0$ one has

$$\Sigma(z^{\alpha},\overline{z^{\beta}}) = z^0\overline{z^2} + z^1\overline{z^3} + z^2\overline{z^0} + z^3\overline{z^1}\ , \tag{14}$$

the standard flat space twistor norm.

H-space

Penrose and Atiyah [3] have shown that the solutions to the good-cut equation for sufficiently small σ^{o}, which are holomorphic on $S^2_{\mathbb{C}}$, form a four-complex dimensional manifold, which is referred to as H-space. The solution can thus be written as

$$Z = Z(\zeta,\tilde{\zeta};z^a) \tag{15}$$

with the four variables z^a parametrizing the space of solutions or acting as the local coordinates of the H-space.

With no attempt at proofs we will now summarize some of the remarkable properties of an H-space.

1) An H-space has a complex Riemannian metric naturally induced on it by

$$ds^2 = g_{ab}(z^a)dz^a dz^b = 8\pi[\oint \frac{dS}{U^2}]^{-1} \tag{16}$$

with

$$dS = \frac{2}{i}\frac{d\zeta\wedge d\tilde{\zeta}}{(1+\zeta\tilde{\zeta})^2}, \tag{17}$$

the area element on the complex unit sphere, and

$$U \equiv dZ = \frac{\partial Z}{\partial z^a}dz^a = Z,_a dz^a, \tag{18}$$

the integration being taken over the real sphere or, to avoid singularities, an appropriate deformation of the real sphere into the complex. We wish to emphasize that it is not at all obvious that the integral in (16) defines a quadratic metric. It does however follow from the properties of the good-cut equation. (The metric obtained from (16) with the solution in equation (6) is the Minkowski metric.)

2) An H-space is Ricci flat, i.e.

$$R_{ab} = 0. \tag{19}$$

3) The Weyl tensor of an H-space is self-dual, i.e.

$$C^{\ *}_{abcd} = iC_{abcd} \tag{20}$$

In other words for each real asymptotically flat physical space-time (and hence for each σ^o), we can generate via the good-cut equation a complex self-dual solution of the Einstein equations.

If the conjugate good-cut equation is used, one obtains in a similar manner an $\tilde{H}$-space with an anti-self-dual Weyl tensor.

Remembering that the starting point of this investigation was the search for asymptotically shear free null surfaces in an asymptotically flat space-time, it is natural to inquire into the asymptotic flatness of the H-space and to see if one can construct the H-space of an H-space. It turns out that not all H-spaces are asymptotically flat--a necessary but by no means sufficient condition for asymptotic flatness being

$$\sigma^o(u,\zeta,\tilde{\zeta}) \to 0 \text{ as } u \to \infty.$$

If however we assume that an H-space is asymptotically flat then one has these further results:

4) The radiation field of the H-space (i.e. the asymptotic behavior of the H-space Weyl tensor) is identical to that of the original physical space.

5) The H-space constructed from an H-space is the original H-space.

6) The $\tilde{H}$-space of an H-space (or the H-space of an $\tilde{H}$-space) is complex Minkowski space.

As a final point we wish to briefly discuss a version [6] (due largely to G. Sparling [7]) of the Atiyah-Ward approach to self or anti-self dual Yang-Mills theory which closely resembles the H-space construction.

Consider an (n × n)-matrix-valued holomorphic function $A(u,\zeta,\tilde{\zeta})$ on $S^2_{\mathbb{C}}$ with $u = x^a \ell_a(\zeta,\tilde{\zeta})$ (see Eq. 8), where the variables x^a are to be interpreted as standard Minkowski space coordinates; and consider the following differential equation for the matrix $G(x^a,\zeta,\tilde{\zeta})$:

$$(1+\zeta\tilde{\zeta}) \frac{\partial}{\partial\zeta} G = -GA \ . \tag{21}$$

We claim (with no proof given here) that a regular solution (regular in the sense of holomorphic in $S^2_{\mathbb{C}}$) generates a self-dual GL(n,$\mathbb{C}$) Yang-Mills field and that all such fields are so generated. Specifically the vector potential is given in terms of G by

$$A_a(x^a) = G,_a G^{-1} + \eth h \ell_a - h \eth \ell_a \tag{22}$$

with

$$h = [\,(1+\zeta\tilde{\zeta}) \frac{\partial}{\partial\tilde{\zeta}} (G,_a G^{-1})\,] \ell^a \tag{23}$$

$$\eth h = (1+\zeta\tilde{\zeta})^2 \frac{\partial}{\partial\zeta} \frac{h}{(1+\zeta\tilde{\zeta})}$$

$$\eth \ell_a = (1+\zeta\tilde{\zeta}) \frac{\partial}{\partial\zeta} \ell_a .$$

Though it is not obvious, one can show that A_a is independent of ζ and $\tilde{\zeta}$ and yields a self-dual Yang-Mills field.

It is Eq. (21) that is the analogue for Yang-Mills theory of the good-cut equation.

Appendix A

If cross-sections of an arbitrary line bundle over S^2 are represented by homogeneous functions of π^A and $\bar{\pi}^{A'}$ (A = 0,1) such that $f(\lambda\pi^A, \bar{\lambda}\bar{\pi}^{A'}) = \lambda^{w-s}\bar{\lambda}^{w+s} f(\pi^A, \bar{\pi}^{A'})$, then the spin-weight s and conformal-weight w function $\eta_{(s,w)}(\zeta,\bar{\zeta})$ is defined [8,9] by

$$\eta_{(s,w)}(\zeta,\bar{\zeta}) = f(\zeta,1,\bar{\zeta},1)P^{-w}$$

$$P = \frac{1}{2}(1+\zeta\bar{\zeta}). \tag{A.1}$$

Using the homogeneity of f one also has

$$f(\pi^A, \bar{\pi}^{A'}) = (\pi^1)^{w-s}(\bar{\pi}^{1'})^{w+s} P^w \eta_{(s,w)}(\zeta,\bar{\zeta})$$

$$\zeta = \frac{\pi^0}{\pi^1}, \quad \bar{\zeta} = \frac{\bar{\pi}^{0'}}{\bar{\pi}^{1'}}. \tag{A.2}$$

These bundles classified by their homogeneity degrees or by s and w, though distinct from the point of view of representation theory, are not topologically distinct. In fact the first Chern class which characterizes (topologically) any smooth line bundle over S^2 is given by

$$c = -2s. \tag{A.3}$$

Appendix B

We will here list a series of problems and questions for which we do not know the answers.

1) To me the most important question is what relationship (if any) does the theory described here have with physics? Penrose [1] argues very

convincingly that an H-space should be viewed as a non-linear graviton, though how to make a quantum theory of gravity from this idea still eludes us. In addition H-space theory appears to be remarkably well adapted to the theory of equations of motion [10,11], though again we are at an impasse.

2) We would like a method of producing solutions to the good cut equation. Though G. Sparling [12] has produced a finite set of solutions and R. Lind [13] has invented a formalism to produce another set, these are quite limited both in number and interesting qualities. One can on the other hand produce [14,15] solutions to the self-dual vacuum Einstein equations with considerable ease, though the general solution is not known.

3) We would like a good useful definition of an asymptotically flat H-space. The definition in current use [10] is rather awkard and unattractive.

4) What are the conditions on the σ° in the good cut equation which guarantee an asymptotically flat H-space?

5) Given [6] an $A(u,\zeta,\tilde{\zeta})$, how can one construct regular solutions to (21)?

6) What are the conditions on A such that the Yang-Mills field is "asymptotically flat?"

7) Are all Ricci flat, self-dual spaces in some local sense equivalent to H-spaces, i.e. are they derivable from a good cut equation? If so, how does one find the σ°?

References

1 R. Penrose, Gen. Rel. Grav. 7(1976), 31.

2 E. T. Newman, J. Porter, K. P. Tod, Gen. Rel. Grav. to appear.

3 R. Hansen, E. T. Newman, R. Penrose, K. P. Tod, to be published, Proc. Roy. Soc.

4 M. Ko, E. T. Newman, K. P. Tod, in "Asymptotic Structure of Space-Time" ed. P. Esposito and L. Witten, Plenum Press, (1976).

5 M. Ko, E. T. Newman, R. Penrose, J. Math. Phys. 18, 58(1977).

6 E. T. Newman, On Source Free Yang-Mills Theories to appear in Phys. Rev.

7 G. Sparling, Preprint Univ. of Pgh.

8 W. D. Curtis, D. E. Lerner, J. Math. Phys. 19, 874(1978).

9 A. Held, E. T. Newman, R. Posadas, J. Math. Phys. 11, 3145(1970).

10 M. Ko, M. Ludvigsen, E. T. Newman, K. P. Tod, The Theory of H-space, preprint, University of Pittsburgh.

11 M. Ludvigsen, G. R. G. 8, 357(1977).

12 G. Sparling, K. P. Tod, Preprint, University of Pittsburgh.

13 Private correspondence

14 J. Plebenski, J. Math. Phys. 16, 2395(1975).

15 A. Janis, W. Fette, E. T. Newman, J. Math. Phys. 17, 660(1976).

Acknowledgment

Research supported by the National Science Foundation.

E T NEWMAN
Department of Physics
University of Pittsburgh
Pittsburgh, Pennsylvania 15260

K P Tod

Remarks on asymptotically flat $\mathcal{H}$-spaces

1. INTRODUCTION

In this article, I wish to discuss the notion of asymptotic flatness for a complex solution of Einstein's equations. I shall be concerned only with left-flat space-times, where the Weyl tensor is self-dual, which arise as H-spaces [1-4].

To define a complex null infinity $\mathbb{C}\mathcal{I}$ for a complex space-time, one might seek a structure analogous to the $\mathbb{C}\mathcal{I}$ of Minkowski space. However, nothing as "big" as this can be expected because a holomorphic metric with curvature will necessarily have singularities which, in particular, will occur at null infinity. Further, a definition in terms of a boundary will not work since $\mathbb{C}\mathcal{I}$, being six real dimensional, is not a boundary of the original eight real dimensional manifold. (This is related to the fact that the null cone does not separate the tangent space; there is no definition of "time-like" or "space-like" directions in a complex space-time.) There is also the possibility that a number of different non-singular regions may be added as limit points, resulting in a number of possible asymptotes or parts of $\mathbb{C}\mathcal{I}$. However, when dealing with an H-space, there are extra structures which can be employed in a definition of asymptotic flatness to avoid these pitfalls.

I shall begin by briefly recalling the salient points of the H-space construction and giving some motivation for why one cares about asymptotic flatness. The method of development is to proceed naively, introducing what seems necessary into the definition. Then with the aid of an example in §3, it is possible to polish up the definition. The H-space construction, [1-4], begins with the real future null infinity $\mathbb{R}\mathcal{I}^+_M$ of an asymptotically flat space-time M and thickens it slightly into the complex, obtaining $\mathbb{C}\mathcal{I}^+_M$. Geometric quantities on $\mathbb{R}\mathcal{I}^+_M$, notably the asymptotic shear $\sigma^o(u,\zeta,\bar{\zeta})$ which determines the asymptotic gravitational radiation field in M, are assumed to be analytic, so that they extend to holomorphic functions on $\mathbb{C}\mathcal{I}^+_M$. This effectively restricts the distance into the complex which the thickening of $\mathbb{R}\mathcal{I}^+_M$ can go. The H-space of M [1] is defined to be the four complex dimensional manifold of "good cuts" of $\mathbb{C}\mathcal{I}^+_M$, that is of regular solutions to the "good cut equation"

$$\eth^2 Z(\zeta,\tilde{\zeta}) = \sigma^o(Z(\zeta,\tilde{\zeta}),\zeta,\tilde{\zeta}). \tag{1.1}$$

Here $\eth$ is roughly $\frac{\partial}{\partial\zeta}$ and $\tilde{\zeta}$ is the complexification of $\bar{\zeta}$, a complex coordinate independent of $\bar{\zeta}$. Regularity means that the solutions $Z(\zeta,\tilde{\zeta})$ are non-singular on the Riemann sphere of ζ when $\tilde{\zeta} = \bar{\zeta}$. (See [1-4] or the article of Newman in this volume for further details).

We write the complete solution of (1.1) as $Z(z^a,\zeta,\tilde{\zeta})$ where the z^a are four parameters on which the solution depends i.e. they are coordinates on the H-space. It is possible to define a quadratic metric and curvature tensor for H-space in terms of the "good cut function" Z. The curvature satisfies the Einstein equations automatically and further is self-dual (or left-flat) [1-4]. Thus σ^o, the data for the original real solution, also determines a left-flat, complex solution. A picture of the H-space construction as a projection of the self-dual part of the radiation field of the original real space-time suggests itself. For this to be appropriate, we need to know that the construction is idempotent i.e. that the H-space construction applied to an H-space reproduces that H-space. Since the construction is only defined on asymptotically flat space-times, we therefore need to know when the H-space is asymptotically flat.

In §2, following remarks of R. Penrose, I point out that there is a canonical identification between $\mathbb{C}\mathcal{I}^+_{\mathrm{M}}$ and what one would want to call $\mathbb{C}\mathcal{I}_H$, the null infinity of H-space. Thus one knows what $\mathbb{C}\mathcal{I}_H$ is, the question remaining is whether it can be smoothly attached to H-space as a set of limit points. Because of this identification, one also knows what the "real" points of $\mathbb{C}\mathcal{I}_H$ are, namely the ones corresponding to $\mathbb{R}\mathcal{I}^+_{\mathrm{M}}$. Thus $\mathbb{C}\mathcal{I}_H$ appears as a thickening of a real $\mathbb{R}\mathcal{I}^+_H$. The definition of asymptotic flatness for H-space therefore begins with the assumption that there is a "real slab" (a term due to M. Ludvigsen) which is the complex thickening of a "real slice" and on which the metric is non-singular. The metric itself is *not* real on the real slice. The real slice is simply defined by a reality condition on the H-space coordinates and is a real submanifold which should extend to infinity. The real slice can be moved around inside the real slab.

For $\mathbb{C}\mathcal{I}_H$ to appear as a thickening of the boundary of the real slice, we need to ensure that null geodesics from the real slab escape to infinity. One would not expect all null geodesics from a point in the real slab to escape to infinity, since some of them will necessarily leave the slab and may encounter singularities. Instead, the second stage in the definition is to assume that null geodesics in directions "close to being real" get away. Specifically, we demand that in the $S^2 \times S^2$ of possible null direction at a point there is an anti-holomorphic "diagonal" S^2 of null geodesics escaping to infinity. (Anti-holomorphic for the following reason: if η is a stereographic coordinate on the first S^2 labelling primed spinors $\pi_{A'}$, and $\tilde{\eta}$ labels unprimed spinors λ_A on the second S^2 then we want a relation $\bar{\eta} = f(\tilde{\eta})$ to play the rôle of complex conjugation of spinors in a real space-time. This then defines the analogue of real null directions.) We shall see in §3 that this is still a little stronger than necessary. In an H-space there is a preferred affine parameter on null geodesics, defined up to an additive constant. Consequently one can consider moving real affine parameter distances along a null geodesic, and we need only demand that these real directions avoid singularities and get to infinity.

With these conditions on an H-space (and one other which is probably redundant) it is possible to prove the idempotence of the H-space construction [3,6]. The analogous $\tilde{H}$-space construction, projecting the anti-self-dual part of the radiation field, is defined from the analytic extension of the complex conjugate of (1.1). Then the $\tilde{H}$-space of an asymptotically flat H-space is simply flat i.e. complex Minkowski space. These relations may be written symbolically [3]

$$H^2 = H; \ \tilde{H}\, H = 0 = H\, \tilde{H}.$$

The content of this introduction and §2 is drawn from many conversations with the relativity groups of the University of Pittsburgh and the Mathematical Institute, Oxford. The identification of $\mathbb{C}I_H$ with $\mathbb{C}I^+_{\mathrm{M}}$ in particular is mainly due to R. Penrose.

The exact solution of §3 appears in [11] and the null geodesic equations for this solution have been solved in collaboration with E.T. Newman and W. Threatt.

Responsibility for any errors and the vaguer speculations is my own.

§2. THE IDENTIFICATION OF $\mathbb{C}I_H$

We show how the $\mathbb{C}I_H$ of an H-space is identified with the $\mathbb{C}I^+_{\mathrm{M}}$ of the space-time M from which it arises with the aid of a series of figures showing $\mathbb{C}I_H$, asymptotic projective (dual) twistor space $\mathbb{P}T^*$ [4,7,8] and H-space itself.

Figure 1a shows $\mathbb{C}I^+_{\mathrm{M}}$ with a good cut Z. Figure 1b is the familiar cube

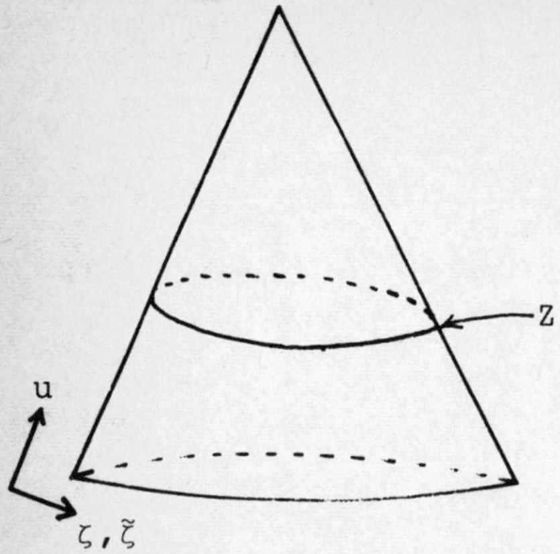

1a: $\mathbb{C}\mathscr{I}^+_M$ with a good cut Z

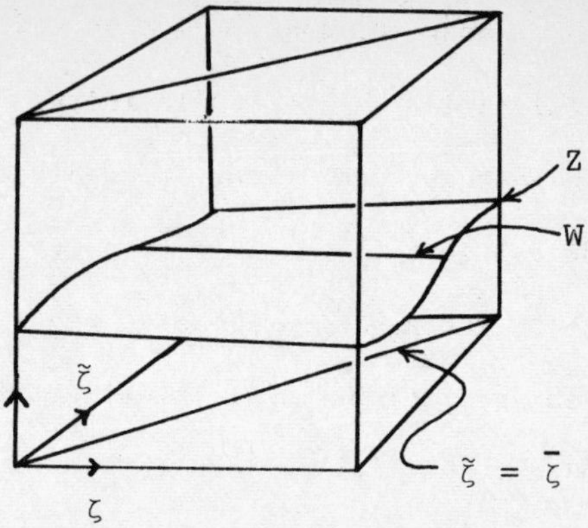

1b: $\mathbb{C}\mathscr{I}^+_M$ showing the real diagonal and a twistor line $W_\bullet$ lying on Z

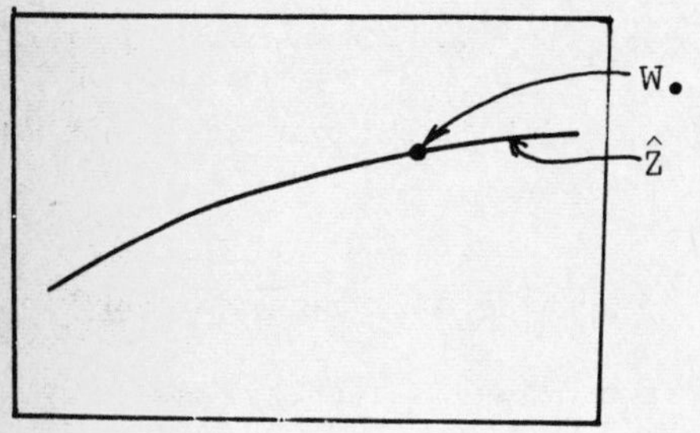

1c: $\mathbb{P}\mathfrak{T}^*$ with the compact holomorphic curve $\hat{Z}$ passing through W.

1d: The point z of $\mathcal{H}$-space corresponding to the good cut Z

Figure 1: The relationships between $C\mathscr{I}^+_M$, $\mathbb{P}\mathfrak{T}^*$ and $\mathcal{H}$-space.

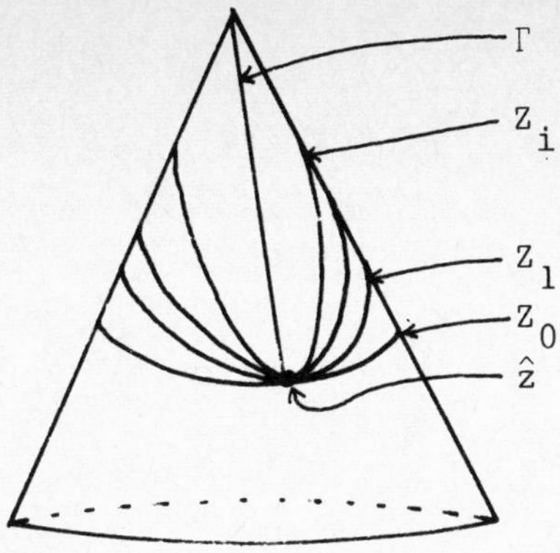

2a: The good cuts Z_i tipping up towards the generator Γ

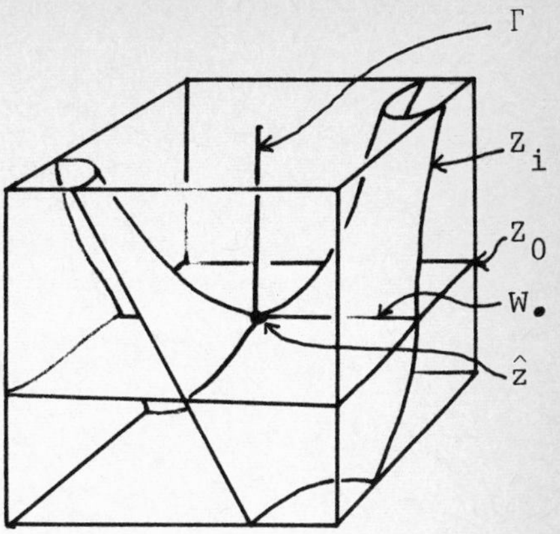

2b: A typical Z_i meeting Z_0 in $W_\bullet$ with tangency at $\hat{z}$

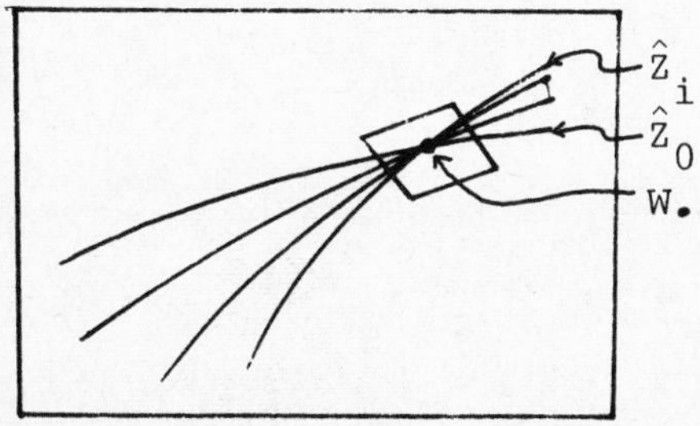

2c: The curves $\hat{Z}_i$ through $W_\bullet$ lying in a 2-plane element

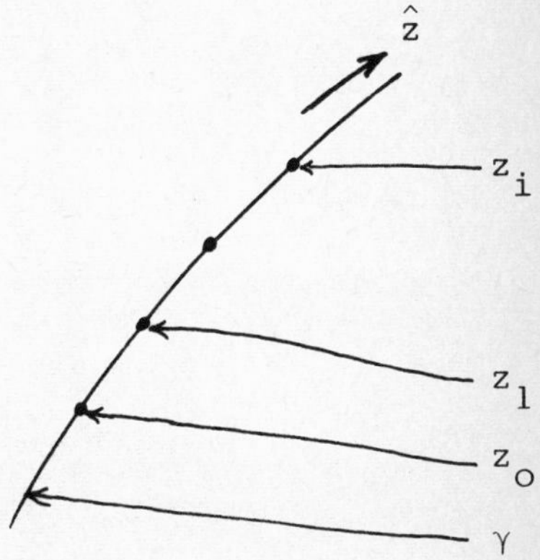

2d: The sequence of points z_i on the geodesic γ in $\mathcal{H}$-space.

Figure 2: A null geodesic in $\mathcal{H}$-space from the various points of view.

picture of $\mathbb{C}\mathcal{I}^+_M$ [4,7,8] showing the good cut Z ruled by geodesics or "twistor lines" in the ζ-direction, and indicating the real diagonal, $\tilde{\zeta} = \bar{\zeta}$, where $\mathbb{C}\mathcal{I}^+_M$ meets $\mathbb{R}\mathcal{I}^+_M$. Figure 1c is $\mathbb{P}\mathcal{T}^*$; the points of $\mathbb{P}\mathcal{T}^*$ are the twistor lines of figure 1b and the good cut Z appears as a compact holomorphic curve $\hat{Z}$, topologically S^2, in $\mathbb{P}\mathcal{T}^*$. Finally, figure 1d shows $\mathcal{H}$-space where the good cut Z represents a point z.

Two points of $\mathcal{H}$ are null separated if the corresponding curves intersect in $\mathbb{P}\mathcal{T}^*$ [2,4,5] or equivalently, if the corresponding good cuts have a whole twistor line in common. There will then be a point on the common twistor line where the two good cuts are actually tangent. A null geodesic γ through a point z_o in $\mathcal{H}$-space now corresponds to the set of all good cuts which intersect the good cut Z_o corresponding to z_o in the same twistor line W, with the same point of tangency, $\hat{z}$. Moving to points z_i at larger and larger affine distances from z_o along γ has the effect of tipping up the good cuts Z_i more and more, as shown in figure 2. Figure 2a shows the Z_i tipping up and figure 2 b shows one typical Z_i meeting Z_o in W_o with tangency at $\hat{z}$. The behaviour of Z_i on the diagonal in figure 2b corresponds to figure 2a. In figure 2c, the curves $\hat{Z}_i$ are shown meeting W_o and lying in a 2-plane element defined by the ζ-coordinate of $\hat{z}$. The tipped up good cuts approach the degenerate good cut which is simply the generator Γ of $\mathbb{C}\mathcal{I}^+_M$ leading up from $\hat{z}$. Thus the "point at infinity" on γ may be canonically identified with the point $\hat{z}$ on $\mathbb{C}\mathcal{I}^+_M$.

If the point $\hat{z}$ has coordinates $(u_o,\zeta_o,\tilde{\zeta}_o)$ then the points z of $\mathcal{H}$-space whose corresponding good cuts pass through $\hat{z}$ have coordinates z^a satisfying

$$Z(z^a,\zeta_o,\tilde{\zeta}_o) = u_o \qquad (2.1)$$

For fixed $(u_o,\zeta_o,\tilde{\zeta}_o)$, (2.1) is the equation of a null hypersurface Σ in $\mathcal{H}$-space (since the $\mathcal{H}$-space gradient $Z_{,a}$ of Z at fixed ζ and $\tilde{\zeta}$ is null [2,4]).

Thus all the geodesics γ which are regarded as having the same point at infinity $\hat{z}$ lie on the null hypersurface Σ, and this construction is seen to mimic the "ideal points" construction of $\mathbb{R}\mathcal{I}^+_M$ [9].

The significant question now is whether $\mathbb{C}\mathcal{I}_H$ is suitably attached to H-space, or equivalently whether the tipped up good cuts approach suitably smoothly to the degenerate good cut Γ. Since the tipped up good cuts will reach larger and larger values of $|u|$, this is a question about the behaviour of $\sigma^o(u,\zeta,\tilde{\zeta})$ for large $|u|$. Exactly what condition on σ^o is necessary for asymptotic flatness of the H-space is still unclear. The example of §3 has

$$\sigma^o \sim |u|^{-3}\ , \text{ large } |u| \tag{2.2}$$

and seems to be asymptotically flat in a reasonable sense. However, (2.2) is stronger than one might expect for a general M. In particular, there are reasons for expecting that σ^o will not even vanish for both large positive and large negative u in a single Bondi frame. (This has also been pointed out by B.D. Bramson [13].) Typically though, there might be two Bondi frames related by a supertranslation in one of which σ^o behaves like (2.2) for large positive u and in the other for large negative u.

A stronger condition than (2.2), which has arisen in conversation with M. Ludvigsen, E.T. Newmann and G.A.J. Sparling, is

$$\sigma^o(u,\zeta,\tilde{\zeta}) \sim \sum_{n\geq 2} \frac{A_n(\zeta,\tilde{\zeta})}{u^{n+1}} \tag{2.3}$$

where $\eth^n A_n = 0$

Thus $A_n = \sum_{2\leq \ell\leq n} a_n^{\ell m}\ {}_2Y_{\ell m}(\zeta,\tilde{\zeta})$

so that the higher spherical harmonics in σ^o fall off more rapidly as $|u| \to \infty$. Such a σ^o has a number of interesting properties (not all of which have been proved!):

a) it is invariant under the Poincaré subgroup of the BMS group,

b) it is the condition for the H-space to be regular at i^+,

c) if σ^o is used in the Kirchhoff integral [10] to obtain a spin-two zero-rest-mass field ψ_{ABCD} on Minkowski space then the null data of ψ_{ABCD} reproduces σ^o. Equivalently, ψ_{ABCD} is regular at i^+ or the twistor function defining ψ_{ABCD} is analytic in the neighbourhood of the line at infinity $I^{\alpha\beta}$ in twistor space.

It would be of some interest to know whether such a σ^o is likely to arise in a physically reasonable space-time.

§3. AN EXAMPLE OF AN H-SPACE

It was first noticed by G.A.J. Sparling [12] that the good cut equation (1.1) was solvable with

$$\sigma^o(u,\zeta,\tilde{\zeta}) = \frac{\lambda f(\tilde{\zeta})}{(u+ia)^3(1+\zeta\tilde{\zeta})^2} \tag{3.1}$$

where f is a quartic polynomial, λ is an arbitrary complex scale factor and a is an arbitrary non-zero real constant included to make σ^o non-singular on $\mathbb{R}I^+_M$. The specific case when f = 1 is dealt with at some length in [11]. This example is sufficiently simple that the sorts of question raised in §§1 and 2 are tractable. The solution may be summarised as follows: the good cut function $Z(z^a,\zeta,\tilde{\zeta})$ is given by

$$(Z + ia)^2 = (z + ia)^2 + \lambda s^2 \tag{3.2}$$

$$\text{where } z = z^a\, l_a(\zeta,\tilde{\zeta}) \tag{3.3a}$$

$$s = s^a\, l_a(\zeta,\tilde{\zeta}) \tag{3.3b}$$

$$l_a(\zeta,\tilde{\zeta}) = \frac{1}{1+\zeta\tilde{\zeta}}(1,\zeta,\tilde{\zeta},\zeta\tilde{\zeta}) \tag{3.3c}$$

$$z^a = (u,X,Y,v) \tag{3.3d}$$

$$s^a = \frac{1}{\Delta}(Y,v,0,0) \tag{3.3e}$$

$$\Delta = (u + ia)(v + ia) - XY \tag{3.3f}$$

so that (u,X,Y,v) are the four complex parameters on which the solution depends.

As $\lambda \to 0$ the solution becomes

$$Z = z$$

which gives Minkowski space for the $\mathcal{H}$-space with the usual coordinates

$$\begin{aligned} u &= t - z & v &= t + z \\ X &= x + iy & Y &= x - iy. \end{aligned} \tag{3.4}$$

The metric for non-zero λ is

$$ds^2 = 2dudv - 2dXdY - \frac{2\lambda}{\Delta^3}(Ydv-(v+ia)dY)^2. \tag{3.5}$$

Since the metric is in Kerr-Schild form there is a canonical flat background and one expects the curvature to be algebraically special. In fact, the curvature is type N, non-singular away from $\Delta = 0$ and has no singularities as a function of λ (being actually proportional to λ). Regarded as a linearised solution on Minkowski space, (3.5) is an elementary state [7] corresponding to an imploding then exploding gravitational wave, everywhere non-singular on the real Minkowski space. In particular, it is asymptotically flat which encourages the belief that it should still be asymptotically flat in the non-linear theory.

One problem with (3.2) is that one must take a square root to obtain Z explicitly and Z will not be a regular function on the sphere $\tilde{\zeta} = \bar{\zeta}$ unless simple zeroes of the right hand side are avoided (i.e. no cuts). This is reasonable, since a zero of the right hand side corresponds to a singularity in σ^{o}. Factorising (3.2) as

$$(Z + ia)^2 = H\tilde{H} \tag{3.6a}$$

$$H = z + ia + i\lambda^{\frac{1}{2}} s \tag{3.6b}$$

$$\tilde{H} = z + ia - i\lambda^{\frac{1}{2}} s \tag{3.6c}$$

we see that simple zeroes of H or $\tilde{H}$ are to be avoided. It turns out that, with (u,X,Y,v) as in (3.4) and (t,x,y,z) all real, H and $\tilde{H}$ are non-zero on the sphere $\tilde{\zeta} = \bar{\zeta}$ provided

$$|\lambda| < a^2. \tag{3.7}$$

With (3.7) holding therefore, there is a canonical real slice in the $\mathcal{H}$-space for which the good cuts are well-defined. This can be thickened into the complex to provide the real slab of §1 since Δ will be non-zero there, so the metric and curvature are non-singular.

Note that, although the metric has no singularities as a function of λ, the actual region of $\mathbb{C}^4$ for which the good cuts are well-defined does depend on λ. If (3.7) is satisfied, the good cuts exist in a thickened neighbourhood of the real slice in the canonical flat background. Larger values of λ will have good cuts existing in a thickened neighbourhood of a translation of this real slice by an imaginary time-like vector.

The next step is to solve the null geodesic equation for the metric (3.5). In currently unpublished work of Newman, Threatt and myself this has been done and remarkably it is possible to solve the equations completely although the metric has only one Killing vector. There are two simplifying features of the general $\mathcal{H}$-space which aid this calculation. Firstly, the gradient $Z_{,a}$ of the good cut function Z at fixed ζ and $\tilde{\zeta}$ is null, so that at a point $Z_{,a}$

sweeps out the entire null cone as ζ and $\tilde{\zeta}$ vary. This provides a scaling for all null vectors and thus a preferred affine parameter on all null geodesics. Secondly the gradient $Z_{,a}$ at fixed ζ and $\tilde{\zeta}$ is actually geodesic:

$$g^{ab}\, Z_{,a}\, Z_{;bc} = 0$$

where ; is the H-space covariant derivative. Thus a null geodesic $z^a(\tau)$ through the point $z^a(0)$ in the direction of $g^{ab}\, Z_{,b}(z^c(0),\zeta,\tilde{\zeta})$ is a solution of the first order equations

$$\frac{d}{d\tau}\, z^a(\tau) = g^{ab}\, Z_{,b}(z^c(\tau),\zeta,\tilde{\zeta})$$

i.e. there are always four first integrals of the null geodesic equation in an H-space and τ is the preferred affine parameter, defined up to additive constants. One may now define "real" null directions as the ones given by $Z_{,a}$ when $\tilde{\zeta} = \bar{\zeta}$. This provides the anti-holomorphic S^2 of null directions for the definition of §1. We now need to know whether, by going real affine distances in real null directions from points on the real slice, we encounter the singularity or escape to infinity. This requires Δ as a function of τ, initial position $z^a(0) = z^a_0$ and initial direction $\dot{z}^a_0 = g^{ab}\, Z_{,b}(z^c_0,\zeta_o,\bar{\zeta}_o)$, which is given by

$$\Delta^2 = (Z_o + ia)^2\, (\tau-\alpha)(\tau-\beta) \tag{3.8}$$

where
$$\begin{aligned}
Z_o &= Z(z^a_o,\zeta_o,\bar{\zeta}_o)\\
\alpha &= -\Delta_o\, H_o^{\frac{1}{2}}\, \tilde{H}_o^{-3/2}\\
\beta &= -\Delta_o\, \tilde{H}_o^{\frac{1}{2}}\, H_o^{-3/2}\\
\Delta_o &= \Delta(z^a_o)\\
H_o &= H(z^a_o,\zeta_o,\bar{\zeta}_o)\\
\tilde{H}_o &= \tilde{H}(z^a_o,\zeta_o,\bar{\zeta}_o)\ .
\end{aligned}$$

Thus every null geodesic has two points α,β in the complex τ-plane where Δ vanishes and the curvature is singular. The question is whether, with the reality conditions, these points can ever be on the real τ-axis. Unfortunately this question is at present unresolved but the indications are that α and β are necessarily in the lower half τ-plane *provided* (3.7) holds. If this turns out to be true, then the H-space *is* asymptotically flat. Further, if both singularities are in the lower half τ-plane we may take this as a definition for the H-space to correspond to a positive frequency non-linear graviton, [5]. It was conjectured in [11] that this solution would be positive frequency in a reasonable sense, on the basis of the shear σ^o being positive frequency in u and the linearised spin-two zero-rest-mass field corresponding to (3.5) being positive frequency in Minkowski space.

In conclusion, we have an apparently rather coordinate dependent notion of asymptotic flatness for a left flat space arising from the H-space construction. The definition leans heavily on the good cut function $Z(z^a,\zeta,\tilde{\zeta})$ to define the real slice, real null directions and the affine parameter on null geodesics, which may be thought of as "memories" of the original real space-time from which the H-space arose. It seems quite appropriate to use this structure and it is gratifying that there is then an example other than flat space which (modulo a proof that $\mathrm{Im}\ \alpha < 0$, $\mathrm{Im}\ \beta < 0$) satisfies the conditions of the definition.

REFERENCES

1. E.T. Newman in "General Relativity and Gravitation" ed. G. Shaviv and J. Rosen, Wiley, New York 1975.

2. M. Ko, E.T. Newman and K.P. Tod in "Asymptotic Structure of Spacetime" ed. F.P. Esposito and L. Witten, Plenum Press, New York 1976.

3. M. Ko, M. Ludvigsen, E.T. Newman and K.P. Tod, "The Theory of H-space" submitted to Phys. Rep.

4. R. O. Hansen, E.T. Newman, R. Penrose and K.P. Tod "The Metric and Curvature Properties of H-space", Proc.Roy.Soc. (Lond.) to appear.

5. R. Penrose, G.R.G. 7 (1976) 31.

6. M. Ludvigsen, Ph.D. thesis, University of Pittsburgh, 1978.

7. R. Penrose and M.A.H. MacCallum, Phys.Rep. 6C (1973) 242.

8. E.J. Flaherty, "Hermitian and Kahlerian Structures in Relativity" Lecture Notes on Physics 46 Springer-Verlag, Berlin 1976.

9. R. Geroch, E.H. Kronheiner and R. Penrose, Proc. Roy.Soc. (Lond.) A327 (1972) 545

10. E.T.Newman and R. Penrose, Proc.Roy.Soc. (Lond.) A305 (1968) 175

11. G.A.J. Sparling and K.P. Tod, "An Example of an H-space", J.Math.Phys. to appear.

12. G.A.J. Sparling, private communication.

13. B.D. Bramson, private communication.

K.P. Tod
Department of Physics and Astronomy
University of Pittsburgh, Pittsburgh,
Pa 15260
U.S.A. and

The Mathematical Institute
Oxford
England

J Isenberg and P B Yasskin

Twistor description of non-self-dual Yang-Mills fields

§1. INTRODUCTION

Recently there has been an increasing awareness of the physical importance of self-dual Yang-Mills fields (e.g. their contribution to the Feynman path integrals of quantum chromodynamics) [1]. The self-dual condition is, however, a debilitating restriction for certain purposes: (1) Real Yang-Mills fields on a real Minkowski space cannot be self-dual; (2) Self-dual fields cannot solve the Y-M equations with sources. Besides these restrictions, the desire to obtain all of the contributions (no matter how small) to the quantum integrals leads one to consider non-self-dual Yang-Mills fields.

An important tool in the study of Y-M fields has been the Ward [2] correspondence between self-dual Y-M fields over complex-Minkowski space, and certain holomorphic principal G-bundles over projective twistor space. The Ward correspondence, discussed extensively in this volume, has lead to the classification [3] and construction [4] of many self-dual Y-M fields. With an eye towards analogous classification and construction, we have modified the Ward correspondence and obtained a "twistorial representation" for the general non-self-dual Yang-Mills gauge fields [5].

To motivate our modification we recall that the set of totally null complex planes in CM has two connected components: the α-planes, P_α, and the β-planes, P_β. The α-planes are spanned by anti-self-dual bivectors and parametrized by the projective twistors. The β-planes are spanned by self-dual bivectors and parametrized by the projective dual twistors. By using P_α as the base space for its G-bundles, the Ward correspondence produces Y-M fields which are flat over each α-plane--a condition which is equivalent to self-duality. It is therefore clear that the modified correspondence for non-self-dual fields cannot use bundles over P_α. Instead we use bundles over L, the space of all complex null lines in CM. The corresponding gauge fields are flat over each such line, but this is of course no restriction; so we get the general non-self-dual fields.

The properties of L, including its relationship to P_α and P_β, are discussed in §2. The basic correspondence between bundles over L and gauge fields on CM is stated as Theorem 1 in §3 and sketched as a geometric and algebraic construction in §4.

Unlike self-dual gauge fields, general gauge fields do not automatically satisfy the currentless Y-M equations:

$$D{*}F = 0. \tag{1.1}$$

While this fact permits our correspondence (unlike Ward's) to handle fields with a current source, i.e.

$$D{*}F = {*}J \neq 0, \tag{1.2}$$

it is also useful to know necessary and sufficient conditions on the bundle over L such that the corresponding Y-M field does indeed have vanishing current. We state such conditions in the form of Theorem 2, in §3. The conditions are rather complicated and the proof is fairly long. However, one of the purposes of this paper is to present this proof. We do so in §5.

Other restrictions on the gauge fields can be realized as conditions on the bundle over L. In reference 5 we state such conditions for reality, flatness, self-duality, and holonomic commutativity.

Note that the use of L instead of P_α (or P_β) entails a bonus: Since the null lines, in the guise of null geodesics, survive the passage to a general curved spacetime, our correspondence should also survive and produce Yang-Mills fields on a curved background. The null planes do not generally survive this passage and hence the standard Ward correspondence does not work for a general curved spacetime.

§2. LINE SPACE

Line Space, L, is defined to be the space of all null lines in conformally completed complex Minkowski space, CM. We shall often want to discuss fields which are not defined over all of CM, but only over a subset $S \subset CM$. So we define L(S) as the space of null lines which intersect S. Similarly, for $r \in S$, we define L(r) as the space of null lines containing r. These definitions, and similar ones for P_α and P_β, appear in Table 1.

We assume that the reader is familiar with the spinor and twistor parametrizations of the various spaces (our notation appears in Table 2.), and with the geometry of the Penrose correspondence [6] between these spaces (as exemplified by the definitions in Table 1). Of particular importance is the fact that every null line lies in a unique α-plane and in a unique β-plane and is precisely their intersection. This implies the existence of the maps in the diagram,

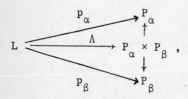

$$\begin{array}{ccc} & \xrightarrow{p_\alpha} & P_\alpha \\ & & \uparrow \\ L & \xrightarrow{\Lambda} & P_\alpha \times P_\beta \\ & & \downarrow \\ & \xrightarrow{p_\beta} & P_\beta \end{array} ,$$

where p_α and p_β are projections and Λ is 1 - 1. The map Λ is not onto. In fact an α-plane, Z, and β-plane, W, intersect iff their twistors satisfy $Z^\alpha W_\alpha = 0$, in which case they intersect in a null line. Hence, Λ imbeds L as the 5-complex dimensional hypersurface, $Z^\alpha W_\alpha = 0$, within $P_\alpha \times P_\beta = CP^3 \times CP^3$.

CM	P_α	P_β	L
r := point	$P_\alpha(r)$:= {α-planes through r} = CP^1	$P_\beta(r)$:= {β-planes through r} = CP^1	L(r) := {null lines through r} = $CP^1 \times CP^1$
Z := totally null plane spanned by anti-self-dual bivector = complex 2-fold	Z := α-plane	$P_\beta(Z)$:= {β-planes intersecting Z} = CP^2	$p_\alpha^{-1}(Z)$:={null lines contained in Z} = CP^2
W := totally null plane spanned by self-dual bivector = complex 2-fold	$P_\alpha(W)$:= {α-planes intersecting W} = CP^2	W := β-plane	$p_\beta^{-1}(W)$:={null lines contained in W} = CP^2
ℓ := null line = complex 1-fold	$p_\alpha(\ell)$:= unique α-plane containing ℓ	$p_\beta(\ell)$:= unique β-plane containing ℓ	ℓ := null line
S := subset of CM	$P_\alpha(S)$:= {α-planes intersecting S}	$P_\beta(S)$:= {null lines intersecting S}	L(S) := {null lines intersecting S}

CM:	vector	$r^a = r^{AA'}$	
CM×CM:	vector pair	(p^a, q^a)	
	or vector pair	(r^a, s^a)	

[Here p^a and q^a are the coordinates on each factor, $r^a = (p^a + q^a)/\sqrt{2}$ is the diagonal coordinate, and $s^a = (p^a - q^a)/\sqrt{2}$ is the off-diagonal coordinate.]

P_α:	projective twistor	Z^α	
	or (spinor, conjugate spinor)	$(\eta^A, \zeta_{A'})$	[up to complex scale]
P_β:	projective dual twistor	W_α	
	or (spinor, conjugate spinor)	$(\omega_A, \mu^{A'})$	[up to complex scale]
$P_\alpha \times P_\beta$:	(proj. twistor, proj. dual twistor)	(Z^α, W_α)	
	or complex 6-tuple on each patch $W^{[m]}$ ...	(δ, x^Σ)	
L:	(proj. twistor, proj. dual twistor)	(Z^α, W_α)	[where $Z^\alpha W_\alpha = 0$]
	or complex 5-tuple on each patch $U^{[m]}$ ...	x^Σ	
B_α:	(vector, conjugate spinor)	$(p^a, \zeta_{A'})$	[up to scale of $\zeta_{A'}$]
B_β:	(vector, spinor)	(q^a, ω_A)	[up to scale of ω_A]
$B_\alpha \times B_\beta$:	(vector, conj. spinor, vector, spinor) ..	$(p^a, \zeta_{A'}, q^a, \omega_A)$	[up to scales of $\zeta_{A'}$ and ω_A]
B_L:	(vector, conjugate spinor, spinor)	$(r^a, \zeta_{A'}, \omega_A)$	[up to scales of $\zeta_{A'}$ and ω_A]

We also emphasize that $L(r) = CP^1 \times CP^1$ is a compact complex manifold. This is crucial to our construction which uses the fact that any holomorphic function on a compact complex manifold must be a constant.

There are certain spaces which are irrelevant to the statement of the theorems, but are crucial to the proofs and also to the explicit statement of the correspondence. These are the flag spaces[7]: B_L, B_α and B_β. An element of B_L consists of a null line in CM and a point on that line. Hence there are projections

$$CM \longleftarrow B_L \xrightarrow{\Pi_L} L. \tag{2.2}$$

There are analogous definitions and projections for B_α and B_β. The flag spaces are related to each other by a diagram similar to (2.1). The coordinate representations of the various maps appear in Table 3.

In the proof of Theorem 2, we need the commutative diagram,

$$\begin{array}{ccccc} CM & \longleftarrow & B_L & \xrightarrow{\Pi_L} & L \\ \Delta \big\downarrow & & \big\downarrow \Delta_B & & \big\downarrow \Lambda \\ CM \times CM & \longleftarrow & B_\alpha \times B_\beta & \xrightarrow{\Pi} & P_\alpha \times P_\beta \end{array} . \tag{2.3}$$

We also use the transition functions for bundles over $L(S)$, $P_\alpha(S) \times P_\beta(S)$, $B_L(S)$ and $B_\alpha(S) \times B_\beta(S)$. So we let $\{W^{[m]}\}$ be a finite open cover of $P_\alpha(S) \times P_\beta(S)$ which is sufficiently refined so that any bundle over $P_\alpha(S) \times P_\beta(S)$ can be specified by its transition functions, $\tilde{g}^{[mn]}$, on $W^{[m]} \cap W^{[n]}$. We then define

$$U^{[m]} := W^{[m]} \cap L \subset L(S), \tag{2.4}$$

$$V^{[m]} := \Pi_L^{-1}(U^{[m]}) \subset B_L(S), \text{ and} \tag{2.5}$$

$$X^{[m]} := \Pi^{-1}(W^{[m]}) \subset B_\alpha(S) \times B_\beta(S). \tag{2.6}$$

TABLE 3: PARAMETERS AND COORDINATES FOR THE MAPS

$$\Pi_L : B_L \to L : (r^{AA'}, \zeta_{A'}, \omega_A) \mapsto (\frac{i}{\sqrt{2}} r^{AA'} \zeta_{A'}, \zeta_{A'}, \omega_A, -\frac{i}{\sqrt{2}} r^{AA'} \omega_A)$$

$$\Pi_\alpha : B_\alpha \to P_\alpha : (p^{AA'}, \zeta_{A'}) \mapsto (ip^{AA'} \zeta_{A'}, \zeta_{A'})$$

$$\Pi_\beta : B_\beta \to P_\beta : (q^{AA'}, \omega_A) \mapsto (\omega_A, -iq^{AA'} \omega_A)$$

$$\Pi : B_\alpha \times B_\beta \to P_\alpha \times P_\beta : (p^{AA'}, \zeta_{A'}, q^{AA'}, \omega_A) \mapsto (ip^{AA'} \zeta_{A'}, \zeta_{A'}, \omega_A, -iq^{AA'} \omega_A)$$

$$\Delta : CM \to CM \times CM : (r^a) \mapsto (r^a/\sqrt{2}, r^a/\sqrt{2})$$

$$\Delta_B : B_L \to B_\alpha \times B_\beta : (r^a, \zeta_{A'}, \omega_A) \mapsto (r^a/\sqrt{2}, \zeta_{A'}, r^a/\sqrt{2}, \omega_A)$$

$$\Lambda : L \to P_\alpha \times P_\beta : (\eta^A, \zeta_{A'}, \omega_A, \mu^{A'}) \mapsto (\eta^A, \zeta_{A'}, \omega_A, \mu^{A'})$$

or

$$\Lambda : U^{[m]} \to W^{[m]} : (x^\Sigma) \mapsto (0, x^\Sigma)$$

Since diagram (2.3) commutes, we have

$$V^{[m]} = X^{[m]} \cap B_L. \tag{2.7}$$

Finally, we need coordinates on $P_\alpha(S) \times P_\beta(S)$ which are adapted to $L(S)$. Thus on each neighborhood, $W^{[m]}$, we use coordinates (δ, x^Σ), $\Sigma = 1,\ldots,5$, where $U^{[m]}$ is the $\delta = 0$ surface and the (x^Σ) are coordinates on $U^{[m]}$.

§3. THEOREMS

Using the language and notation of §2, we now state our theorems. The first of these describes the basic correspondence:

Theorem 1 (Correspondence):

Let G be a complex Lie subgroup of GL(n,C). Let S be an open subset of the conformal completion of CM whose intersection with each null line is either empty or connected and simply connected. There is a 1-1 correspondence between

(1) holomorphic principal G-bundles E_S over S with holomorphic connection A and curvature F; and

(2) holomorphic principal G-bundles $E_{L(S)}$ over L(S) which have trivial restriction $E_{L(r)}$ to L(r) for all $r \in S$.

In §4, we describe how to build the line space bundle $E_{L(S)}$, given the spacetime bundle E_S with its gauge field A and F; and also how to construct E_S, A and F, given $E_{L(S)}$. The proof [8], that these constructions are valid and are inverses of each other, is analogous to that of Ward's theorem.

Our other theorem states the conditions on $E_{L(S)}$ which are necessary and sufficient for the gauge field to satisfy the currentless Yang-Mills equations:

Theorem 2 (Currentless):

Let $E_{L(S)}$ be the bundle over L(S) which corresponds (according to Theorem 1) to the gauge field, A_L and F_L, on the bundle E_S over S. Let $\tilde{g}_L^{[mn]}$ be the transition functions for $E_{L(S)}$ appropriate to the cover $U^{[m]}$ of L(S). Then

$$D_L * F_L = 0,$$

iff there exists a G-valued function $\tilde{g}^{[mn]}$ on some neighborhood of $U^{[m]} \cap U^{[n]}$ within each intersection, $W^{[m]} \cap W^{[n]} \subset P_\alpha \times P_\beta$, such that

(a) $\tilde{g}^{[mn]}\big|_L = \tilde{g}_L^{[mn]}$ and

(b) $\tilde{g}^{[mn]}\tilde{g}^{[nk]} = \tilde{g}^{[mk]} + \mathcal{O}(\delta^4)$.

This theorem (proven in §5) says that the currentless Yang-Mills equations are satisfied iff the bundle $E_{L(S)}$ has a third order extension from L(S) to some neighborhood of L(S) within $P_\alpha(S) \times P_\beta(S)$. If the bundle can be extended to all orders, then we get the much stronger restriction that the Yang-Mills field also satisfies the holonomic commutativity condition [9]. In the case of SU(2), this condition implies that the gauge field is self-dual, anti-self-dual, or abelian.

§4. CONSTRUCTION OF THE CORRESPONDENCE IN THEOREM 1

The Correspondence Theorem implies that, using a given line space bundle $E_{L(S)}$, one may construct a spacetime bundle E_S with a connection A_L and curvature F_L; and vice versa. We now sketch how this is done.

The crucial fact in our construction, as well as Ward's, comes from complex analysis: Any holomorphic function from a compact, complex manifold into C must be a constant. We assume G is a complex Lie subgroup of GL(n,C). Then any holomorphic cross section of a trivial principal G-bundle over a compact, complex manifold is determined by its value at a single point. Further, the set of such cross sections may be identified with the fibre at that point.

Given the bundle $E_{L(S)}$, the point set of E_S is constructed as follows: For each $r \in S$, let the fibre over r be the set $\pi^{-1}(r)$ of holomorphic cross sections of $E_{L(r)}$, and let $E_S = \bigcup_{r \in S} \pi^{-1}(r)$. Since $L(r) = CP^1 \times CP^1$ is compact and $E_{L(r)}$ is trivial, $\pi^{-1}(r)$ may be identified with G. We complete the structure of E_S by constructing a global trivialization in the form of a global cross section: Let the bundle $E_{BL(S)}$ be the pull-back of $E_{L(S)}$ along Π_L: $B_L(S) \to L(S)$. Since $E_{BL(S)}$ is trivial, choose a global cross section. This section determines a holomorphic cross section of $E_{L(r)}$ for each r in S and hence a global cross section of E_S.

The connection is constructed by first defining a transport T along null directions: Assume r and $r' \in S$ lie on a null line ℓ. Each $\psi \in \pi^{-1}(r')$ is a holomorphic cross section of $E_{L(r')}$. Since every holomorphic cross section of $E_{L(r)}$ is determined by its value at a point, we define the transport of ψ from r' to r along ℓ as the unique holomorphic cross section $T_{r \leftarrow r'}\psi$ of $E_{L(r)}$, whose value at $\ell \in L(r) \cap L(r')$ is $(T_{r \leftarrow r'}\psi)(\ell) = \psi(\ell)$. See figure 1. This transport rule defines a covariant derivative provided it is linear in the differentiating direction. We check this by finding a formula for the connection 1-form A_L as follows:

If $\tilde{g}_L^{[mn]}$ are the transition functions of $E_{L(S)}$ on $U^{[m]} \cap U^{[n]}$, then

$$g_L^{[mn]} = \tilde{g}_L^{[mn]} \circ \Pi_L \tag{4.1}$$

are the transition functions of $E_{BL(S)}$ on $V^{[m]} \cap V^{[n]}$. Since $E_{BL(S)}$ is trivial, its transition functions split:

$$g_L^{[mn]} = g_L^{[m]}(g_L^{[n]})^{-1}, \tag{4.2}$$

FIGURE 1

$T_{r\leftarrow r'}\psi$ IS THE TRANSPORT

OF ψ FROM r to r' ALONG ℓ

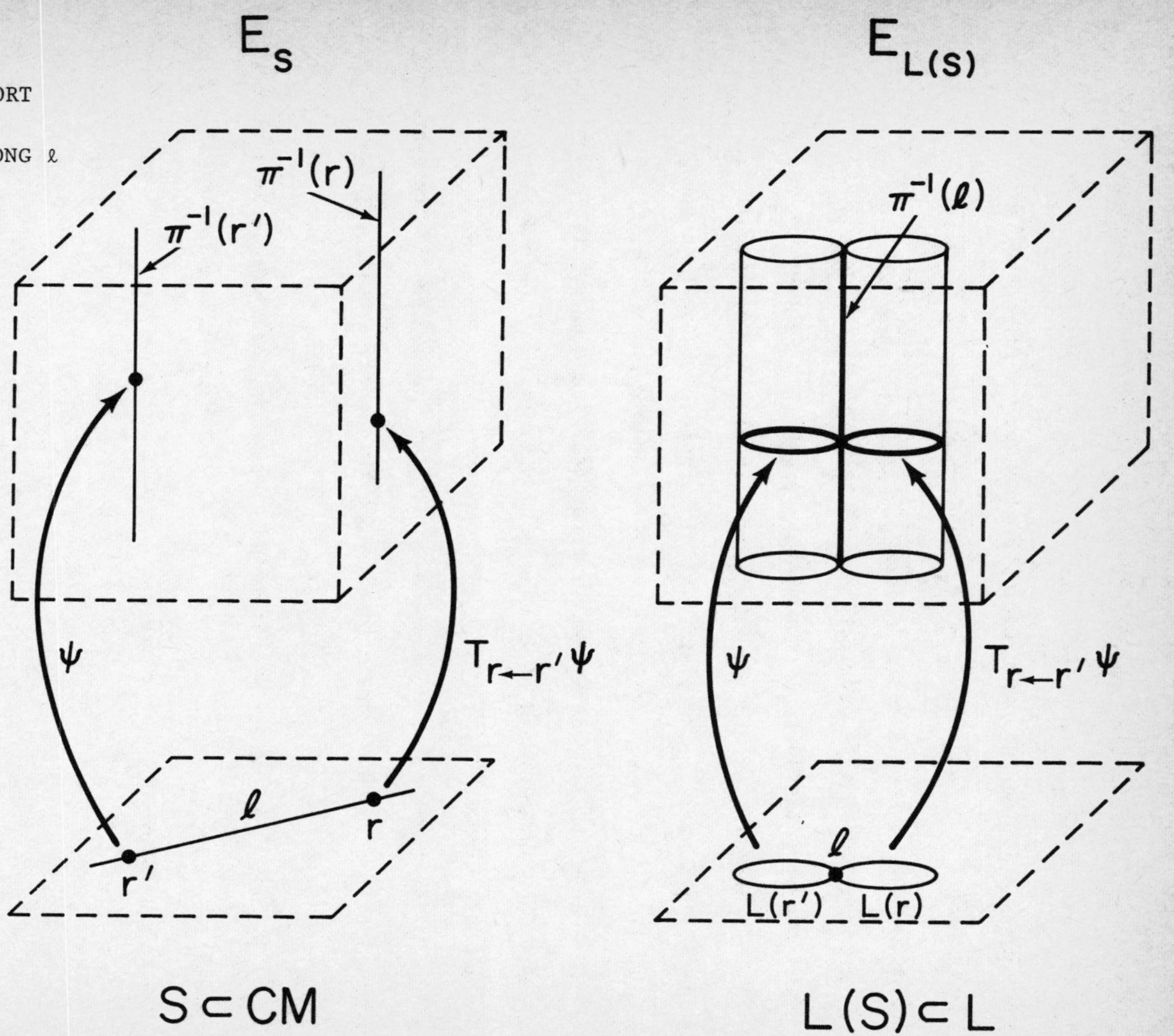

where $g_L^{[m]}$ is defined on $V^{[m]}$. (Note: The choice of splitting corresponds to the choice of global cross section of $E_{BL(S)}$ and to the choice of gauge on E_S.) Using these $g_L^{[m]}$'s we find that [10]

$$(A_L)_{\omega\zeta} = (g_L^{[m]})^{-1} \frac{\partial}{\partial r^{\omega\zeta}} g_L^{[m]} \tag{4.3}$$

is the desired 1-form. (We prove that $(A_L)_{BB'}$ from (4.3) is independent of patch and independent of ω and ζ by using

$$\frac{\partial}{\partial r^{\omega\zeta}} g_L^{[mn]} = 0, \tag{4.4}$$

which in turn follows from (4.1).)

The inverse construction--$E_{L(S)}$ from a given E_S with connection A_L--again starts with the specification of the point set (of $E_{L(S)}$): For each $\ell \in L(S)$, let the fibre over ℓ be the set $\pi^{-1}(\ell)$ of covariantly constant cross sections of E_ℓ, the restriction of E_S to ℓ; and let $E_{L(S)} = \bigcup_{\ell \in L(S)} \pi^{-1}(\ell)$. Since such a cross section is determined by its value at a point, $\pi^{-1}(\ell)$ may be identified with G. We complete the structure of $E_{L(S)}$ by constructing a local trivialization for each patch $U^{[m]}$ in the form of a map $\phi_L^{[m]}: \pi^{-1}(U^{[m]}) \rightarrow G$: A global trivialization of E_S provides a map $\phi_S: E_S \rightarrow G$. Pick a holomorphic function, $r^{[m]}: U^{[m]} \rightarrow S$, such that the point $r^{[m]}(\ell)$ lies on the line ℓ. Each $\psi \in \pi^{-1}(\ell)$ is a covariantly constant cross section of E_ℓ which may be evaluated at $r^{[m]}(\ell)$. Thus we define

$$\phi_L^{[m]}(\psi) = \phi_S(\psi[r^{[m]}(\pi[\psi])]). \tag{4.5}$$

It follows that the transition functions are

$$\tilde{g}_L^{[mn]}(\ell) = p \exp\{-\int_{r^{[n]}(\ell)}^{r^{[m]}(\ell)} A_\mu dx^\mu\}, \tag{4.6}$$

where "p" indicates that the exponential integral is path-ordered.

It can be shown [8] that these two constructions are inverses.

§5. PROOF OF THEOREM 2

Our proof of Theorem 2 is motivated by a correspondence between certain Y-M fields on S × S and bundles over $P_\alpha \times P_\beta$. We show that a chain of five statements are equivalent. The first and last appear in the theorem. The intervening conditions concern the extension of a bundle with connection from S to S × S and the extension of a bundle from $B_L(S)$ to $B_\alpha(S) \times B_\beta(S)$. (See diagram (2.3).) The five conditions are as follows:

[I] $D_L * F_L = 0.$ (I.1)

[II] There exists a connection, $\underline{A} = A_{AA'}dr^{AA'} + B_{AA'}ds^{AA'}$, on $E_{S\times S}$ such that

(a) $\Delta^*(\underline{A}) = A_L$, or equivalently,

$$A_{AA'}\Big|_{\Delta(S)} = (A_L)_{AA'}; \text{ and} \tag{II.1}$$

(b) if we define

$$D_{AA'} := \frac{\partial}{\partial r^{AA'}} + A_{AA'} \text{ and} \tag{II.2}$$

$$\nabla_{AA'} := \frac{\partial}{\partial s^{AA'}} + B_{AA'}, \tag{II.3}$$

then we have

$$[\nabla_{AA'},\nabla_{BB'}] = [D_{AA'},D_{BB'}] + O(s^2) \text{ and} \tag{II.4}$$

$$[D_{AA'},\nabla_{BB'}] = i*[D_{AA'},D_{BB'}] + O(s^2). \tag{II.5}$$

(Note: $D_{AA'}$ and $\nabla_{AA'}$ are not really covariant derivatives and equations (II.4) and (II.5) are not really tensor equations.)

[III] There exists a connection, $\underline{A} = P_{AA'}dp^{AA'} + Q_{AA'}dq^{AA'}$, on $E_{S\times S}$ such that

(a) $\Delta^*(\underline{A}) = A_L$, or equivalently,

$$\frac{1}{\sqrt{2}}(P_{AA'} + Q_{AA'})\Big|_{\Delta(S)} = (A_L)_{AA'}; \text{ and} \tag{III.1}$$

(b) if we define

$$d_{AA'} := \frac{\partial}{\partial p^{AA'}} + P_{AA'} \text{ and} \tag{III.2}$$

$$\delta_{AA'} := \frac{\partial}{\partial q^{AA'}} + Q_{AA'}, \tag{III.3}$$

then for all $\zeta^{A'}$ and ω^{A} we have [10]

$$[d_{A\zeta}, d_{B\zeta}] = \mathcal{O}(s^2), \tag{III.4}$$

$$[\delta_{\omega A'}, \delta_{\omega B'}] = \mathcal{O}(s^2), \text{ and} \tag{III.5}$$

$$[d_{AA'}, \delta_{BB'}] = \mathcal{O}(s^2). \tag{III.6}$$

[IV] There exists a G-valued function $g^{[m]}$ on some neighborhood of $V^{[m]}$ within each patch $X^{[m]} \subset B_\alpha \times B_\beta$, and there exists a G-valued function $g^{[mn]}$ on some neighborhood of $V^{[m]} \cap V^{[n]}$ within each intersection $X^{[m]} \cap X^{[n]}$, such that

(a) $$g^{[m]}\Big|_{B_L} = g_L^{[m]}, \tag{IV.1}$$

(b) $$g^{[mn]} = g^{[m]}(g^{[n]})^{-1} + \mathcal{O}(s^4), \text{ and} \tag{IV.2}$$

(c) $$\frac{\partial}{\partial p^{A\zeta}} g^{[mn]} = 0 \text{ and} \tag{IV.3}$$

$$\frac{\partial}{\partial q^{\omega A'}} g^{[mn]} = 0. \tag{IV.4}$$

[V] There exists a G-valued function $\tilde{g}^{[mn]}$ on some neighborhood of $U^{[m]} \cap U^{[n]}$ within each intersection, $W^{[m]} \cap W^{[n]} \subset P_\alpha \times P_\beta$, such that

(a) $$\tilde{g}^{[mn]}\Big|_L = \tilde{g}_L^{[mn]} \quad \text{and} \tag{V.1}$$

(b) $$\tilde{g}^{[mn]}\tilde{g}^{[nk]} = \tilde{g}^{[mk]} + O(\delta^4). \tag{V.2}$$

We now proceed to prove successively the equivalence of conditions [I] through [V].

<u>Transition [A]: [I] <—>[II]</u>

←) We first assume the existence of a connection $\underline{A} = A_{AA'}dr^{AA'} + B_{AA'}ds^{AA'}$ satisfying conditions (II.1)-(II.5). We expand $A_{AA'}$ and $B_{AA'}$:

$$A_{AA'} = \sum_{k=0}^{\infty} A_{AA'kB_1B_1'\dots B_kB_k'}\, s^{B_1B_1'}\dots s^{B_kB_k'}, \tag{A.1}$$

$$B_{AA'} = \sum_{k=0}^{\infty} B_{AA'kB_1B_1'\dots B_kB_k'}\, s^{B_1B_1'}\dots s^{B_kB_k'}. \tag{A.2}$$

Equation (II.1) implies that

$$A_{AA'0} = (A_L)_{AA'}. \tag{A.3}$$

It can be shown that one can always pick a gauge, consistent with (A.3), in which the totally symmetric part of $B_{AA'kB_1B_1'\dots B_kB_k'}$ vanishes for all k; i.e.

$$B_{(AA'kB_1B_1'\dots B_kB_k')} = 0. \tag{A.4}$$

In particular,

$$B_{AA'0} = 0. \tag{A.5}$$

Using (A.1) and (A.2) and this choice of gauge, $[D_{AA'},D_{BB'}]$, $[\nabla_{AA'},\nabla_{BB'}]$ and $[D_{AA'},\nabla_{BB'}]$ can be expanded in $s^{AA'}$. We see that (II.4) and (II.5) imply $D_L * F_L = 0$. We also find that

$$A_{AA'1BB'} = i(*F_L)_{BB'AA'}, \tag{A.6}$$

$$B_{AA'1BB'} = \frac{1}{2}(F_L)_{BB'AA'}, \tag{A.7}$$

$$A_{AA'2BB'CC'} = \frac{1}{4}[(D_L)_{BB'}(F_L)_{CC'AA'} + (D_L)_{CC'}(F_L)_{BB'AA'}], \tag{A.8}$$

and

$$B_{AA'2BB'CC'} = \frac{i}{6}[(D_L)_{BB'}(*F_L)_{CC'AA'} + (D_L)_{CC'}(*F_L)_{BB'AA'}], \tag{A.9}$$

where

$$(*F_L)_{AA'BB'} := \frac{1}{2}\,\varepsilon_{AA'BB'}{}^{CC'DD'}(F_L)_{CC'DD'}. \tag{A.10}$$

(Notice that $D_L *F_L = 0$, (A.6), (A.7) and (A.9) actually follow from the weaker assumptions

$$[\nabla,\nabla] = [D,D] + \mathcal{O}(s^2) \text{ and } [D,\nabla] = i*[D,D] + \mathcal{O}(s^1); \tag{A.11}$$

while $D_L * F_L = 0$, (A.6), (A.7) and (A.8) follow from

$$[\nabla,\nabla] = [D,D] + \mathcal{O}(s^1) \text{ and } [D,\nabla] = i*[D,D] + \mathcal{O}(s^2).) \tag{A.12}$$

$\rightarrow$) Conversely, we assume A_L and F_L satisfy $D_L * F_L = 0$. In some gauge for the bundle $E_{S\times S}$ (specified by some global trivialization) we specify $\underline{A} = A_{AA'}dr^{AA'} + B_{AA'}ds^{AA'}$ by giving $A_{AA'}$ and $B_{AA'}$ as the quadratic polynomials

$$A_{AA'} = A_{AA'0} + A_{AA'1BB'}s^{BB'} + A_{AA'2BB'CC'}s^{BB'}s^{CC'}, \tag{A.13}$$

and

$$B_{AA'} = B_{AA'0} + B_{AA'1BB'}s^{BB'} + B_{AA'2BB'CC'}s^{BB'}s^{CC'}, \tag{A.14}$$

with coefficients given by (A.3), (A.5), (A.6), (A.7), (A.8) and (A.9). Equation (A.3) guarantees that (II.1) is satisfied. We define $D_{AA'}$ and $\nabla_{AA'}$ as in (II.2) and (II.3). Then, if we substitute (A.13) and (A.14) into the exapnsions for $[D_{AA'}, D_{BB'}]$, $[\nabla_{AA'}, \nabla_{BB'}]$ and $[D_{AA'}, \nabla_{BB'}]$, we find that (II.4) and (II.5) are satisfied.

Transition [B]: [II] $\leftrightarrow$ [III]

The same connection $\underline{A}$ appears in conditions [II] and [III]. The coefficients $P_{AA'}$ and $Q_{AA'}$ may be related to $A_{AA'}$ and $B_{AA'}$ by using the coordinate transformation

$$r^{AA'} = \frac{1}{\sqrt{2}}(p^{AA'} + q^{AA'}), \qquad p^{AA'} = \frac{1}{\sqrt{2}}(r^{AA'} + s^{AA'}),$$
$$s^{AA'} = \frac{1}{\sqrt{2}}(p^{AA'} - q^{AA'}), \qquad q^{AA'} = \frac{1}{\sqrt{2}}(r^{AA'} - s^{AA'}). \tag{B.1}$$

Thus we find that

$$A_{AA'} = \frac{1}{\sqrt{2}}(P_{AA'} + Q_{AA'}), \qquad P_{AA'} = \frac{1}{\sqrt{2}}(A_{AA'} + B_{AA'}),$$
$$B_{AA'} = \frac{1}{\sqrt{2}}(P_{AA'} - Q_{AA'}), \qquad Q_{AA'} = \frac{1}{\sqrt{2}}(A_{AA'} - B_{AA'}). \tag{B.2}$$

Consequently, we have

$$D_{AA'} = \frac{1}{\sqrt{2}}(d_{AA'} + \delta_{AA'}), \qquad d_{AA'} = \frac{1}{\sqrt{2}}(D_{AA'} + \nabla_{AA'}),$$
$$\nabla_{AA'} = \frac{1}{\sqrt{2}}(d_{AA'} - \delta_{AA'}), \qquad \delta_{AA'} = \frac{1}{\sqrt{2}}(D_{AA'} - \nabla_{AA'}). \tag{B.3}$$

The first of equations (B.2) shows that equation (II.1) is equivalent to equation (III.1). Using equations (B.3) we relate the commutators:

$$2[D_{AA'},D_{BB'}] = [d_{AA'},d_{BB'}] + [\delta_{AA'},\delta_{BB'}] + [d_{AA'},\delta_{BB'}] + [\delta_{AA'},d_{BB'}], \tag{B.4}$$

$$2[\nabla_{AA'},\nabla_{BB'}] = [d_{AA'},d_{BB'}] + [\delta_{AA'},\delta_{BB'}] - [d_{AA'},\delta_{BB'}] - [\delta_{AA'},d_{BB'}], \tag{B.5}$$

$$2[D_{AA'},\nabla_{BB'}] = [d_{AA'},d_{BB'}] - [\delta_{AA'},\delta_{BB'}] - [d_{AA'},\delta_{BB'}] + [\delta_{AA'},d_{BB'}], \tag{B.6}$$

$$2[d_{AA'},d_{BB'}] = [D_{AA'},D_{BB'}] + [\nabla_{AA'},\nabla_{BB'}] + [D_{AA'},\nabla_{BB'}] + [\nabla_{AA'},D_{BB'}], \tag{B.7}$$

$$2[\delta_{AA'},\delta_{BB'}] = [D_{AA'},D_{BB'}] + [\nabla_{AA'},\nabla_{BB'}] - [D_{AA'},\nabla_{BB'}] - [\nabla_{AA'},D_{BB'}], \tag{B.8}$$

$$2[d_{AA'},\delta_{BB'}] = [D_{AA'},D_{BB'}] - [\nabla_{AA'},\nabla_{BB'}] - [D_{AA'},\nabla_{BB'}] + [\nabla_{AA'},D_{BB'}]. \tag{B.9}$$

It follows that equation (II.4) and (II.5) are equivalent to

$$[d_{AA'},d_{BB'}] = i*[d_{AA'},d_{BB'}] + \mathcal{O}(s^2), \tag{B.10}$$

$$[\delta_{AA'},\delta_{BB'}] = -i*[\delta_{AA'},\delta_{BB'}] + \mathcal{O}(s^2), \text{ and} \tag{B.11}$$

$$[d_{AA'},\delta_{BB'}] = \mathcal{O}(s^2). \tag{B.12}$$

Equation (B.12) is just (III.6). It remains to prove that (B.10) is equivalent to (III.4) and that (B.11) is equivalent to (III.5). To see this, we write

$$[d_{AA'},d_{BB'}] =: \Phi_{AB}\varepsilon_{A'B'} + \Psi_{A'B'}\varepsilon_{AB} \text{ and} \tag{B.13}$$

$$[\delta_{AA'},\delta_{BB'}] =: \Lambda_{AB}\varepsilon_{A'B'} + \Omega_{A'B'}\varepsilon_{AB}. \tag{B.14}$$

(This expansion can be carried out for any antisymmetric tensor.) Using

$$i\,\varepsilon_{AA'BB'}{}^{CC'DD'} = \delta_A^D\delta_B^C\delta_{A'}^{C'}\delta_{B'}^{D'} - \delta_A^C\delta_B^D\delta_{A'}^{D'}\delta_{B'}^{C'}, \tag{B.15}$$

along with the definition of "*", (A.10), we find that

$$\begin{aligned} i{*}[d_{AA'},d_{BB'}] &= \tfrac{i}{2}\varepsilon_{AA'BB'}{}^{CC'DD'}[d_{CC'},d_{DD'}] \\ &= \Phi_{AB}\varepsilon_{A'B'} - \Psi_{A'B'}\varepsilon_{AB} \text{ and} \end{aligned} \tag{B.16}$$

$$i{*}[\delta_{AA'},\delta_{BB'}] = \Lambda_{AB}\varepsilon_{A'B'} - \Omega_{A'B'}\varepsilon_{AB}. \tag{B.17}$$

Consequently, (B.10) and (III.4) are both equivalent to

$$\Psi_{A'B'} = \mathcal{O}(s^2), \tag{B.18}$$

while (B.11) and (III.5) are both equivalent to

$$\Lambda_{AB} = \mathcal{O}(s^2). \tag{B.19}$$

<u>Transition [C]:[III] <->[IV]</u>

$\rightarrow$) We first assume the existence of a connection $\underline{A} = P_{AA'}dp^{AA'} + Q_{AA'}dq^{AA'}$ satisfying conditions (III.1)-(III.6). We expand $P_{AA'}$ and $Q_{AA'}$ in the coordinate $s^{AA'}$, obtaining

$$P_{AA'} = \sum_{k=0}^{\infty} P_{AA'kB_1B_1'\ldots B_kB_k'}\, s^{B_1B_1'}\ldots s^{B_kB_k'}, \tag{C.1}$$

$$Q_{AA'} = \sum_{k=0}^{\infty} Q_{AA'kB_1B_1'\ldots B_kB_k'}\, s^{B_1B_1'}\ldots s^{B_kB_k'}. \tag{C.2}$$

The functions $g^{[m]}$ and $g^{[mn]}$, as yet unknown, may also be expanded:

$$g^{[m]} = \sum_{k=0}^{\infty} g^{[m]}_{kA_1A_1'\ldots A_kA_k'}\, s^{A_1A_1'}\ldots s^{A_kA_k'}, \tag{C.3}$$

$$g^{[mn]} = \sum_{k=0}^{\infty} g^{[mn]}_{kA_1A'_1\ldots A_kA'_k} s^{A_1A'_1}\ldots s^{A_kA'_k}. \tag{C.4}$$

We specify $g^{[m]}$ and $g^{[mn]}$ by specifying the expansion coefficients in (C.3) and (C.4). To guarantee that (IV.1) is satisfied, we require that

$$g^{[m]}_o = g^{[m]}_L. \tag{C.5}$$

Expressions for $g^{[m]}_{1AA'}$, $g^{[m]}_{2AA'BB'}$, $g^{[m]}_{3A\zeta BB'CC'}$ and $g^{[m]}_{3\omega A'BB'CC'}$ may be read out of the expansion of the equations

$$\frac{\partial}{\partial p^{A\zeta}} g^{[m]} = g^{[m]} P_{A\zeta} + \mathcal{O}(s^3) \text{ and} \tag{C.6}$$

$$\frac{\partial}{\partial q^{\omega A'}} g^{[m]} = g^{[m]} Q_{\omega A'} + \mathcal{O}(s^3), \tag{C.7}$$

provided certain consistency conditions are met. These conditions are guaranteed by (III.1), (III.4), (III.5) and (III.6). We now have all the components of $g^{[m]}_{3AA'BB'CC'}$ except $g^{[m]}_{3\omega^\perp\zeta^\perp\omega^\perp\zeta^\perp\omega^\perp\zeta^\perp}$, where $\zeta^\perp$ and $\omega^\perp$ are spinors such that $\zeta^\perp_{A'}\ \zeta^{A'} = 1$ and $\omega^\perp_A\ \omega^A = 1$. For later purposes, it is necessary to require that

$$\frac{\partial}{\partial r^{\omega\zeta}} g^{[m]}_{3\omega^\perp\zeta^\perp\omega^\perp\zeta^\perp\omega^\perp\zeta^\perp} = 0. \tag{C.8}$$

Other than (C.8), $g^{[m]}_{3\omega^\perp\zeta^\perp\omega^\perp\zeta^\perp\omega^\perp\zeta^\perp}$ is arbitrary. Hence we set

$$g^{[m]}_{3\omega^\perp\zeta^\perp\omega^\perp\zeta^\perp\omega^\perp\zeta^\perp} = 0. \tag{C.9}$$

The higher order terms of $g^{[m]}$ are arbitrary provided that (C.3) converges. Hence we set

$$g^{[m]}_{kA_1A'_1\ldots A_kA'_k} = 0, \quad \text{for } k \geq 4. \tag{C.10}$$

This completes the definition of $g^{[m]}$.

To guarantee that (IV.2) is satisfied, we define

$$g^{[mn]}_{kA_1A'_1\ldots A_kA'_k} := \sum_{j=0}^{k} g^{[m]}_{j(A_1A'_1\ldots A_jA'_j} h^{[n]}_{k-jA_{j+1}A'_{j+1}\ldots A_kA'_k)}, \qquad (C.11)$$

for k = 0,1,2,3, where $h^{[n]} = (g^{[n]})^{-1}$. To second order, (IV.3) and (IV.4) are satisfied by virtue of (C.6), (C.7), and (C.11).

We then inductively define the higher order terms in $g^{[mn]}$ in a manner which guarantees that (IV.3) and (IV.4) are satisfied to successively higher orders. In the first step of the induction, consistency is guaranteed by (C.8). At each order, one component of $g^{[mn]}_{kA_1A'_1\ldots A_kA'_k}$ is left undetermined. It is chosen to guarantee consistency at the next order. The induction completes [11] the definition of $g^{[mn]}$ and shows that (IV.3) and (IV.4) are satisfied.

$\leftarrow$) Conversely, we assume the existence of $g^{[m]}$ and $g^{[mn]}$ satisfying (IV.1)-(IV.4). We define a function $f^{[m]}_A(r,s,\zeta,\omega)$, on each patch $X^{[m]} \subset B_\alpha \times B_\beta$, as the quadratic polynomial in $s^{BB'}$ given by the zeroth, first and second order terms in the power series expansion of $(g^{[m]})^{-1} \frac{\partial}{\partial p^{A\zeta}} g^{[m]}$. In order to compare $f^{[m]}_A$ and $f^{[n]}_A$ on $X^{[m]} \cap X^{[n]}$, we substitute (IV.2) into (IV.3) to obtain

$$\frac{\partial}{\partial p^{A\zeta}} [g^{[m]} (g^{[n]})^{-1}] = O(s^3), \qquad (C.12)$$

and hence

$$(g^{[m]})^{-1} \frac{\partial}{\partial p^{A\zeta}} g^{[m]} = (g^{[n]})^{-1} \frac{\partial}{\partial p^{A\zeta}} g^{[n]} + O(s^3). \qquad (C.13)$$

Since $f^{[m]}_A$ and $f^{[n]}_A$ have been defined to be the quadratic parts of each side of (C.13), it follows that

$$f_A^{[m]} = f_A^{[n]} \text{ on } X^{[m]} \cap X^{[n]}. \tag{C.14}$$

Thus there is a global function $f_A(r,s,\zeta,\omega)$ on $B_\alpha \times B_\beta$, consistently defined by

$$f_A(r,s,\zeta,\omega) = f_A^{[m]}(r,s,\zeta,\omega), \tag{C.15}$$

where $(r,s,\zeta,\omega) \in X^{[m]}$.

Since $g^{[m]}$ is homogeneous of degree zero in ζ and ω, the functions $(g^{[m]})^{-1}\frac{\partial}{\partial p^{A\zeta}} g^{[m]}$, $f_A^{[m]}$, and hence f_A, are homogeneous of degree zero in ω and of degree one in ζ. Since f_A is globally defined, there exist functions $P_{AA'}(r,s)$ such that

$$f_A(r,s,\zeta,\omega) =: \zeta^{A'} P_{AA'}(r,s). \tag{C.16}$$

Therefore, for any $(r,s,\zeta,\omega) \in X^{[m]}$,

$$P_{A\zeta}(r,s) = (g^{[m]})^{-1} \frac{\partial}{\partial p^{A\zeta}} g^{[m]} + \mathcal{O}(s^3). \tag{C.17}$$

Similarly, there exist functions $Q_{AA'}(r,s)$ such that

$$Q_{\omega A'}(r,s) = (g^{[m]})^{-1} \frac{\partial}{\partial q^{\omega A'}} g^{[m]} + \mathcal{O}(s^3). \tag{C.18}$$

Using (C.17) and (C.18) we compute

$$\frac{1}{\sqrt{2}} (P_{\omega\zeta} + Q_{\omega\zeta})\Big|_{s=0} = (g^{[m]}\Big|_{s=0})^{-1} \frac{\partial}{\partial r^{\omega\zeta}} (g^{[m]}\Big|_{s=0}). \tag{C.19}$$

Then combining (IV.1) with (C.19) we obtain

$$\frac{1}{\sqrt{2}} (P_{\omega\zeta} + Q_{\omega\zeta})\Big|_{s=0} = (g_L^{[m]})^{-1} \frac{\partial}{\partial r^{\omega\zeta}} g_L^{[m]} = (A_L)_{\omega\zeta}. \tag{C.20}$$

Since ω and ζ are arbitrary, this proves that (III.1) is satisfied.

Next, we define $d_{AA'}$ and $\delta_{AA'}$ according to (III.2) and (III.3). Using (C.17) we compute

$$[d_{A\zeta},d_{B\zeta}] = \frac{\partial}{\partial p^{A\zeta}} P_{B\zeta} - \frac{\partial}{\partial p^{B\zeta}} P_{A\zeta} + [P_{A\zeta},P_{B\zeta}]$$

$$= \frac{\partial}{\partial p^{A\zeta}}[(g^{[m]})^{-1}\frac{\partial}{\partial p^{B\zeta}} g^{[m]}] + (g^{[m]})^{-1} (\frac{\partial}{\partial p^{A\zeta}} g^{[m]})(g^{[m]})^{-1}(\frac{\partial}{\partial p^{B\zeta}} g^{[m]})$$

$$- \frac{\partial}{\partial p^{B\zeta}} [(g^{[m]})^{-1} \frac{\partial}{\partial p^{A\zeta}} g^{[m]}] - (g^{[m]})^{-1}(\frac{\partial}{\partial p^{B\zeta}} g^{[m]})(g^{[m]})^{-1}(\frac{\partial}{\partial p^{A\zeta}} g^{[m]})+\mathcal{O}(s^2)$$

$$= \mathcal{O}(s^2). \qquad (C.21)$$

Similarly, using (C.18) we compute

$$[\delta_{\omega A'},\delta_{\omega B'}] = \mathcal{O}(s^2). \qquad (C.22)$$

Finally, using both (C.17) and (C.18) we compute

$$[d_{A\zeta},\delta_{\omega B'}] = \frac{\partial}{\partial p^{A\zeta}} Q_{\omega B'} - \frac{\partial}{\partial q^{\omega B'}} P_{A\zeta} + [P_{A\zeta},Q_{\omega B'}]$$

$$= \frac{\partial}{\partial p^{A\zeta}} [(g^{[m]})^{-1} \frac{\partial}{\partial q^{\omega B'}} g^{[m]}] + (g^{[m]})^{-1}(\frac{\partial}{\partial p^{A\zeta}} g^{[m]})(g^{[m]})^{-1}(\frac{\partial}{\partial q^{\omega B'}} g^{[m]})$$

$$- \frac{\partial}{\partial q^{\omega B'}} [(g^{[m]})^{-1} \frac{\partial}{\partial p^{A\zeta}} g^{[m]}] - (g^{[m]})^{-1}(\frac{\partial}{\partial q^{\omega B'}} g^{[m]})(g^{[m]})^{-1}(\frac{\partial}{\partial p^{A\zeta}}g^{[m]})+\mathcal{O}(s^2)$$

$$= \mathcal{O}(s^2). \qquad (C.23)$$

For each ζ and ω, the computations (C.21), (C.22), and (C.23) are true for some patch $X^{[m]}$. Hence, (III.4), (III.5), and (III.6) are satisfied.

<u>Transition [D]: [IV] <-> [V]</u>

$\leftarrow$) We first assume the existence of G-valued functions $\tilde{g}^{[mn]}$, on the patches of $P_\alpha \times P_\beta$, satisfying conditions (V.1) and (V.2). Using the projection $\Pi : B_\alpha \times B_\beta \to P_\alpha \times P_\beta$, we pull the functions $\tilde{g}^{[mn]}$ back to

$B_\alpha \times B_\beta$ and define

$$g^{[mn]} = \Pi^* \tilde{g}^{[mn]} = \tilde{g}^{[mn]} \circ \Pi . \tag{D.1}$$

It follows from (D.1) and the coordinate description of Π that, for any spinors λ^A and $\rho^{A'}$, we have

$$g^{[mn]}(p^{AA'},q^{AA'},\zeta_{A'},\omega_A) = g^{[mn]}(p^{AA'}+\lambda^A\zeta^{A'},q^{AA'}+\omega^A\rho^{A'},\zeta_{A'},\omega_A). \tag{D.2}$$

That is, $g^{[mn]}$ is invariant under translations along the α-plane $(p^{AA'},\zeta_{A'})$ and along the β-plane $(q^{AA'},\omega_A)$. Therefore, (IV.3) and (IV.4) follow from

$$\frac{\partial}{\partial p^{A\zeta}} g^{[mn]}(p^{AA'},q^{AA'},\zeta_{A'},\omega_A)$$

$$= \zeta^{A'} \frac{\partial}{\partial p^{AA'}} \tilde{g}^{[mn]}(ip^{BB'}\zeta_{B'},\zeta_{B'},\omega_B,-iq^{BB'}\omega_B) = 0 \text{ and} \tag{D.3}$$

$$\frac{\partial}{\partial q^{\omega A'}} g^{[mn]}(p^{AA'},q^{AA'},\zeta_{A'},\omega_A)$$

$$= \omega^A \frac{\partial}{\partial q^{AA'}} \tilde{g}^{[mn]}(ip^{BB'}\zeta_{B'},\zeta_{B'},\omega_B,-iq^{BB'}\omega_B) = 0.$$

Since the diagram (2.3) commutes, we see that condition (V.1) implies (IV.1):

$$g^{[mn]}\Big|_{B_L} = \tilde{g}^{[mn]} \circ \Pi \circ \Delta_B = \tilde{g}^{[mn]} \circ \Lambda \circ \Pi_L = \tilde{g}_L^{[mn]} \circ \Pi_L = g_L^{[mn]}. \tag{D.4}$$

To prove condition (IV.2) we first show that the $g^{[mn]}$ satisfy the cocycle condition to order s^3:

$$g^{[mn]}g^{[nk]} = g^{[mk]} + O(s^4). \tag{D.5}$$

We prove (D.5) by expanding both $g^{[mn]}$ and $\tilde{g}^{[mn]}$ in power series:

$$g^{[mn]} = \sum_{k=0}^{\infty} g^{[mn]}_{kA_1A'_1\ldots A_kA'_k} s^{A_1A'_1}\ldots s^{A_kA'_k},$$

$$\tilde{g}^{[mn]} = \sum_{k=0}^{\infty} \tilde{g}^{[mn]}_k \delta^k. \tag{D.6}$$

Using the chain rule, we may express the $g^{[mn]}_{kA_1A'_1\ldots A_kA'_k}$ in terms of the $\tilde{g}^{[mn]}_k$. For example,

$$g^{[mn]}_{1AA'} = \frac{\partial g^{[mn]}}{\partial s^{AA'}}\bigg|_{s=0} = \frac{\partial \tilde{g}^{[mn]}}{\partial \delta} \frac{\partial \delta}{\partial s^{AA'}}\bigg|_{s=0} + \frac{\partial \tilde{g}^{[mn]}}{\partial x^{\Sigma}} \frac{\partial x^{\Sigma}}{\partial s^{AA'}}\bigg|_{s=0}$$

$$= \tilde{g}^{[mn]}_1 \frac{\partial \delta}{\partial s^{AA'}}\bigg|_{s=0} + \frac{\partial \tilde{g}^{[mn]}_o}{\partial x^{\Sigma}} \frac{\partial x^{\Sigma}}{\partial s^{AA'}}\bigg|_{s=0} . \tag{D.7}$$

Condition (D.5) then follows directly from the assumption (V.2).

We finally show that the third order splitting condition (IV.2) follows from the third order cocycle condition (D.5) and from the analytic properties of $B_\alpha \times B_\beta$. Recall that the bundle E_{B_L} over B_L is trivial, and so any set of transition functions $g^{[mn]}_L$ for E_{B_L} must split. Therefore, we have

$$g^{[mn]}_o = g^{[mn]}_L = g^{[m]}_o (g_o{}^n)^{-1}. \tag{D.8}$$

Starting with (D.8) and using the fact that the cohomology class

$$H^1(C^4, CP^1 \times CP^1, \mathcal{O}(\text{matrices})) \tag{D.9}$$

vanishes, one can show that the first, second and third order terms in (D.5) imply that $g^{[mn]}_{1AA'}$, $g^{[mn]}_{2AA'BB'}$, and $g^{[mn]}_{3AA'BB'CC'}$ split, in the sense that condition (IV.2) is satisfied.

$\rightarrow$) Conversely, we assume the existence of functions $g^{[m]}$ and $g^{[mn]}$ satisfying (IV.1)-(IV.4). While there is no map from $P_\alpha \times P_\beta$ to $B_\alpha \times B_\beta$ which might be used to pull $g^{[mn]}$ back to $P_\alpha \times P_\beta$, we see that the conditions

(IV.3) and (IV.4) say that the $g^{[mn]}$ satisfy (D.2). Hence, we may define $\tilde{g}^{[mn]}$ by choosing any pair of points $p^{AA'}$ and $q^{AA'}$ such that $p^{AA'}$ is on the α-plane $(\eta^A, \zeta_{A'})$ and $q^{AA'}$ is on the β-plane $(\omega_A, \mu^{A'})$, and then setting

$$\tilde{g}^{[mn]}(\eta^A, \zeta_{A'}, \omega_A, \mu^{A'}) := g^{[mn]}(p^{AA'}, q^{AA'}, \zeta_{A'}, \omega_A). \tag{D.10}$$

This $\tilde{g}^{[mn]}$, of course, satisfies (D.1).

We now consider $\tilde{g}^{[mn]}$ restricted to $L \subset P_\alpha \times P_\beta$. The α-plane and β-plane for any point in L must meet. Therefore, in evaluating $\tilde{g}^{[mn]}$ from (D.10), we may pick $p^{AA'} = q^{AA'}$. Using (D.10), (IV.2), (IV.1) and a relation similar to (D.10) which relates $\tilde{g}_L^{[mn]}$ and $g_L^{[mn]}$, we obtain condition (V.1):

$$\tilde{g}^{[mn]}\Big|_L = g^{[mn]}\Big|_{B_L} = g_L^{[mn]} = \tilde{g}_L^{[mn]}. \tag{D.11}$$

We finally prove that the $\tilde{g}^{[mn]}$ satisfy the third order cocycle condition (V.2). Condition (IV.2) immediately implies (D.5). We again expand $g^{[mn]}$ and $\tilde{g}^{[mn]}$ as in (D.6) and use the chain rule to relate the coefficients as in (D.7). However, we now want the $\tilde{g}_k^{[mn]}$ in terms of the $g^{[mn]}_{kA_1A_1'\ldots A_kA_k'}$ and not vice versa. We can use the chain rule (D.7) only because of the freedom to choose $s^{AA'} = 0$ whenever $\delta = 0$. While we cannot solve (D.7) for the $\tilde{g}_k^{[mn]}$ in terms of only the $g^{[mn]}_{kA_1A_1'\ldots A_kA_k'}$, we can obtain the $\tilde{g}_k^{[mn]}$ in terms of the $g^{[mn]}_{kA_1A_1'\ldots A_kA_k'}$ and the $\tilde{g}_j^{[mn]}$ for $j < k$. Then condition (V.2) follows order by order from (D.5).

Acknowledgments

This work was done in collaboration with Paul Green, whose insight was crucial to the formulation of our theorems.

This work was supported in part by the National Science Foundation under Grant PHY-76-20029.

References

1 R. Jackiw, Rev. Mod. Phys. 49, (1977), 681-706;
S. Coleman, in Erice Lectures - 1977, ed. by A. Zichichi, (Academic Press, NY, to be published);
R. Jackiw, contribution to this volume.

2 R. S. Ward, Phys. Lett. 61A, (1977) 81-82.

3 M. F. Atiyah and R. S. Ward, Commun. Math. Phys. 55, (1977), 117-124.
M. F. Atiyah, N. J. Hitchin and I. M. Singer, Proc. Nat. Acad. Sci. USA 74, (1977), 2662-2663.

4 M. F. Atiyah, N. J. Hitchin, V. G. Drinfeld and Y. I. Manin, Phys. Lett. 65A, (1978), 185-187;
N. H. Christ, E. J. Weinberg, N. K. Stanton, "General Self-Dual Yang-Mills Solutions," 1978, Columbia Univ. preprint CU-TP-119.

5 J. Isenberg, P. B. Yasskin and P. S. Green, Phys. Lett. 78B(1978), 462-464;
E. Witten, Phys. Lett. 77B(1978), 394-398,

6 R. Penrose, in Quantum Gravity, ed. by C. J. Isham, R. Penrose, and D. W. Sciama, (Clarendon, Oxford, 1975), 268-407;
R. O. Hansen and E. T. Newman, Gen. Rel. Grav. 6 (1975), 361-385.

7 The space B_α is the same as the space F_{12} discussed by R. Wells, contribution to this volume.

8 Available upon request.

9 See Theorem 4 in Isenberg, Yasskin and Green, ref. 5.

10 If we regard $g_L^{[m]}$ as functions of $(r^{AA'}, \zeta_{A'}, \omega_A)$, homogeneous of degree 0 in $\zeta_{A'}$ and ω_A, then $\frac{\partial}{\partial r^{\omega\zeta}} := \omega^A \zeta^{A'} \frac{\partial}{\partial r^{AA'}}$, $\frac{\partial}{\partial r^{A\zeta}} := \zeta^{A'} \frac{\partial}{\partial r^{AA'}}$, and $\frac{\partial}{\partial r^{\omega A'}} := \omega^A \frac{\partial}{\partial r^{AA'}}$.

11 Note that we have not yet verified that this series converges.

J ISENBERG and P B YASSKIN
Center for Theoretical Physics
Department of Physics and Astronomy
University of Maryland
College Park, Maryland 20742

E Witten

Some comments on the recent twistor space constructions

In this talk I would like to comment on certain features of the Horrocks-Barth-Atiyah-Hitchin-Manin-Drinfeld (HBAHMD) construction [1] of bundles in $\mathbb{CP}^3$ and self-dual Yang-Mills solutions. (This construction will also be described at this conference by Robin Hartshorne.)

One purpose is to explain in a simple language why this construction works, and incidentally to translate the twistor space construction into Minkowski space language. However, the main motivation for the effort described here is that I feel that the construction as usually presented has a number of features that are too special for broad applicability in physics.

In particular, I would like to de-emphasize those aspects of the construction that are inherently global. Although the treatment of the global Euclidean space problem is very powerful and beautiful, we must remember that physics takes place in Minkowski space and that global considerations are not usually in the forefront in physics. Also, the quantum Yang-Mills field satisfies not the self-dual equation $F_{\mu\nu} = \tilde{F}_{\mu\nu}$ but the second order equation $D^{\mu}F_{\mu\nu} = 0$. If we hope to eventually make applications to quantum field theory, we must probably learn to think about this latter equation in Minkowski space. And this means not a global problem but a local or initial value problem. In particular, it would be very exciting to find an analogue of the HBAHMD construction for the Minkowski space equation $D^{\mu}F_{\mu\nu} = 0$.

With this in mind, I will be describing the HBAHMD construction in a way that is suited for a consideration of the local problem--that is, the problem of finding self-dual Yang-Mills fields that are well defined, not throughout Euclidean space, but just in a small open set thereof. (This corresponds, in $\mathbb{CP}^3$, to a bundle defined not on all of $\mathbb{CP}^3$ but just on a neighborhood of a line.)

First, let us review Ward's construction of the self-dual Yang-Mills solutions [2]. Ward considers (cf. his paper in this volume) the space of all left-handed null two planes α in four dimensional complex Minkowski space $\mathbb{CM}$. Given a self-dual gauge field in $\mathbb{CM}$, Ward introduces a vector bundle on $\mathbb{CP}^3$ as follows. An element of this bundle is a pair (α,ϕ) where α is a null two plane in $\mathbb{CM}$ and ϕ is a (Lorentz) scalar field (in the fundamental representation of the gauge group) that is covariantly constant on α.

We may say that ϕ is defined only on α. But a statement that is more suitable for generalizations is to say that ϕ is defined throughout $\mathbb{CM}$ (rather, throughout as much of $\mathbb{CM}$ as we are working with) but is defined only modulo the addition of a field vanishing on α. According to this definition, in other words, the fiber of Ward's bundle over a given two plane α is the space of all scalar fields ϕ with $D_\mu\phi = 0$ on α, modulo the space of all fields that vanish on α. The bundle E consisting of all pairs (α,ϕ) has, as Ward showed, a natural holomorphic structure, from which the original self-dual Yang-Mills gauge field can be reconstructed.

Now, HBAHMD tell us that this bundle E can be obtained as follows. We introduce vector spaces A, B, and C, and for each point Z^α in $\mathbb{CP}^3$ we introduce linear maps $f(Z):A \to B$, $g(Z):B \to C$. For fixed Z, f and g are linear maps, and also they depend linearly on the homogeneous coordinates Z^α of $\mathbb{CP}^3$. Also, for fixed Z, $g(Z)f(Z) = 0$. The whole picture is

$$A \xrightarrow{f(Z)} B \xrightarrow{g(Z)} C \qquad (1)$$

and the bundle E arises as the kernel of g modulo the image of f. In other words, for fixed Z, the fiber of E at Z is the kernel of g modulo the image of f.

I would like to explain what A,B,C,f, and g are in the construction of reference [1], and why they are that way.

As a first approximation, let us imagine that A, B, and C each are the space of all scalar fields (in the fundamental representation of the gauge group) and let us try to define maps f and g such that E will be the kernel of g divided by the image of f.

We will define f, corresponding to a given Z or to a given two plane α, to be multiplication by a suitably chosen function which vanishes on α. Then im f will consist precisely of scalar fields that vanish on α.

It is a little harder to define g. We will define g so that the kernel of g consists of all scalar fields that are covariantly constant on α. Then ker g/im f will be exactly the space of all covariantly constant fields on α.

The above is a good approximation to what we want, but it suffers from a basic deficiency. We are trying to construct self-dual gauge fields. A self-dual gauge field is completely determined by certain arbitrary functions of three variables, which one can choose, for instance, to be the values of the gauge field on the initial value hypersurface $t = 0$. However, the spaces A, B, and C, as defined above, are spaces of arbitrary functions of four variables. At best, it is extremely redundant to describe the self-dual gauge fields, which really depend only on functions of three variables, in terms of spaces A, B, and C that involve arbitrary functions of four variables.

The problem is not just a problem of redundancy; it is a problem of principle. If our goal is to find a construction along the lines of (1) that could be used, at least in principle, to solve the initial value problem, then A, B, C, f, and g should be spaces, and maps, that are known explicitly once the initial data are given. I think that this is a reasonable property to insist on in a construction like [1].

To arrange that A, B, and C are spaces of functions of three variables, not four, and that they (and the maps f and g between them) are such as to be known explicitly once the initial data are given, we should choose A, B, and C to be spaces, not of arbitrary scalar functions, but of solutions of some auxiliary equations in the background self-dual gauge field. Then A, B, and C will be spaces of certain arbitrary functions of three variables--the initial data of the auxiliary equations.

To proceed further, I must be more specific, and tell you what function that vanishes on α we will use in defining f.

The function we use will be not a scalar but a spinor,

$$\psi_A = c_A + x_{AA'}d^{A'} \tag{2}$$

where $(c_A, d^{A'})$ are constants, the choice of which depends on α, and where $x_{AA'}$ are the coordinates of complex Minkowski space.

Now, you will recognize that $\psi_A = 0$ defines a two dimensional surface which is in fact a left-handed null two plane--a surface corresponding to a twistor or element of $\mathbb{CP}^3$. (In fact, c_A and $d^{A'}$ can be regarded as the homogeneous coordinates of $\mathbb{CP}^3$.)

We will sometimes identify a two plane α with the spinor (2) that vanishes on it.

I will also need the spinors of opposite helicity

$$\tilde{\psi}_{A'} = \tilde{d}_{A'} + x_{AA'}\tilde{c}^A . \tag{3}$$

It is an essential, and not completely obvious, fact that the spinor spaces ψ and $\tilde{\psi}$ are dual to each other. In fact, we can define a bilinear form

$$(\psi, \tilde{\psi}) = \tilde{d}_{A'} d^{A'} - c_A \tilde{c}^A \quad . \tag{4}$$

This form is obviously Lorentz invariant. It is not obvious that it is conformally invariant, but this is, in fact, true.

Now, let us define the spaces A, B, and C.

A will be a space of unprimed spinors u_A that satisfy a certain equation that will be described later.

B will be a space of scalar fields ϕ that satisfy a certain equation that also will be described later.

C will be the space of the unprimed spinors w_A which satisfy the Dirac equation

$$D^A{}_{A'} w_A = 0. \tag{5}$$

Now let us define our maps f and g. Remember, f and g will depend on the choice of two plane in $\mathbb{CM}$ or, equivalently, on the choice of the spinor ψ_A which vanishes on the two plane.

We define $f: A \to B$ by

$$f(\psi) u_A = \psi^A u_A \tag{6}$$

Then the image of f consists of functions that vanish on the two plane $\psi^A = 0$.

To define g I must first state what equation the scalar fields ϕ in B satisfy. This equation will be given in a rather implicit form (but see eq. (13) for a more explicit version) and at first sight looks rather peculiar.

The condition to be satisfied by ϕ is that its covariant derivative has an expansion

$$D_{AA'}\phi = \sum_{a=1}^{4} \tilde{\psi}_{A'}{}^{a} w_{A}{}^{a} \tag{7}$$

where $\tilde{\psi}_{A'}{}^{a}$, a=1, ... ,4 are our four spinors of primed type defined in equation (3), and where $w_{A}{}^{a}$, a=1, ... ,4 are required to be solutions of the Dirac equation, that is, elements of C.

Now we can define our map $g(\psi):B \to C$. We define

$$g(\psi)\phi = \sum_{a=1}^{4} (\psi, \tilde{\psi}^{a}) w_{A}{}^{a}$$

We must show that the kernel of g consists precisely of those scalar functions that are covariantly constant on the two surface on which ψ vanishes.

To argue this, I will proceed in a noncovariant way. Consider the special case $c_A = 0$, $d^{1'} = 1$, $d^{2'} = 0$, that is $\psi_A = x_{A1'}$. Any case could be mapped onto this one by a Lorentz transformation.

Also, I will choose a basis for the primed spinors $\tilde{\psi}_{A'}$:

$$\begin{aligned} &\tilde{\psi}^{(1)}_{1'} = 1, \quad \tilde{\psi}^{(1)}_{2'} = 0 \\ &\tilde{\psi}^{(2)}_{1'} = 0, \quad \tilde{\psi}^{(2)}_{2'} = 1 \\ &\tilde{\psi}^{(3)}_{A'} = x_{1A'} \\ &\tilde{\psi}^{(4)}_{A'} = x_{2A'}. \end{aligned} \tag{9}$$

In this basis the formula $D_{AA'}\phi = \sum_{a} \tilde{\psi}_{A'}{}^{a} w_{A}{}^{a}$ means that

$$D_{A1'}\phi = w_{A}{}^{(1)} + x_{11'}\, w_{A}{}^{(3)} + x_{21'}\, w_{A}{}^{(4)}. \tag{10}$$

Also, $(\psi,\tilde{\psi})$ is nonzero, with our choice of ψ and in this basis, only for $\tilde{\psi}^{(1)}$. Therefore $g(\psi)\phi$, with out definitions, is zero if and only if $w_A^{(1)} = 0$. In this case

$$D_{A1'}\phi = x_{11'}\, w_A^{(3)} + x_{21'}\, w_A^{(4)} \tag{11}$$

so that

$$D_{A1'}\phi = 0 \text{ if } x_{B1'} = 0. \tag{12}$$

This is the desired result, stating that to be in the kernel of $g(\psi)$, ϕ must be covariantly constant on the two plane on which ψ vanishes.

As one can see by differentiating equation (7) and symmetrizing on the primed indices, antisymmetrizing on the unprimed ones (and using the fact that w_A satisfies the Dirac equation while $\psi_{A'}$ satisfies the twistor equation $\partial_{AA'}\tilde{\psi}_{B'} + \partial_{AB'}\tilde{\psi}_{A'} = 0$) to get a consistency condition, fields ϕ that satisfy (7) exist only if the background gauge field with respect to which D_μ is defined is self-dual; it is here that self-duality enters into the construction.

Now let us tie up the loose ends in the definitions of A, B, and C. C was already defined as the space of solutions of the Dirac equation (5). B we define as the space of all scalar fields whose covariant derivatives can be expanded in the form (7), and A we define as the space of spinors u_A such that, for any ψ^A of type (2), $u_A\psi^A$ is an element of B. These definitions are rather indirect and are not immediately transparent. With some thought one can see that a more explicit and equivalent way to define B is to say that B consists of the space of all scalar fields ϕ that can be written

$$\phi = a + x^{AA'} b_{AA'} + x^2 c \tag{13}$$

where a, $b_{AA'}$, and c all satisfy the covariant Laplace equation, $D_\mu D^\mu s = 0$. However, a given element of B can be written in the form (13) in many ways, as a result of which this expansion is awkward to use. The general element of A can be defined in a somewhat similar way.

Since we have shown that the kernel of g consists of scalar functions covariantly constant on α, and the image of f consists of scalar functions vanishing on α, it may seem quite plausible that the kernel of g modulo the image of f is precisely the fiber of Ward's bundle E. What still must be shown is that B is large enough that any desired covariantly constant value on α is assumed by some element of B, and that A is large enough relative to B that the equivalence classes, the kernel of g modulo the image of f, depend only on the value on α. These facts can be established by relatively elementary arguments analogous to those above.

Rather than continuing in this vein, let us shift here to a more formal line of argument. Manin and Drinfeld [1] have given a very elegant derivation of the construction under discussion here using the Kazul complex, which is an exact sequence of sheaves. Let Ω^k be the kth antisymmetric tensor product of the cotangent bundle of $\mathbb{CP}^3$. Then there is an exact sequence of sheaves

$$0 \to \Omega^3(3) \to \Omega^2(2) \to \Omega^1(1) \to \mathbb{C} \to \mathbb{C}|_S \to 0 \tag{14}$$

depending on the arbitrary choice of a point S in $\mathbb{CP}^3$ (and where $\mathbb{C}|_S$ represents the complex numbers sitting at S). Following Manin and Drinfeld, one tensors (14) with E(-1) and writes down the standard long exact sequences of cohomology groups, in order to obtain information about E. In this way one finds that if one defines

$$A = H^1(E \otimes \Omega^2(1))$$

$$B = H^1(E \otimes \Omega) \tag{15}$$

$$C = H^1(E(-1))$$

and f and g as the natural maps induced from (14), then E is in fact ker g/im f.

The point I wish to make here is that one can easily see that global properties are not needed in this argument. If instead of all of $\mathbb{CP}^3$ one is working with a neighborhood of a line in $\mathbb{CP}^3$, the Manin-Drinfeld argument still goes through. (Some of the arguments for why certain cohomology groups vanish are modified, but the net conclusion is the same.) A neighborhood of a line in $\mathbb{CP}^3$ corresponds to a small open set in complex Minkowski space, and thus it corresponds to the problem of a self-dual solution that is defined only in a small open set. Thus, the Manin-Drinfeld derivation shows that the construction that we are discussing here is valid for this local problem as well as for the global problem.

In fact, from the Minkowski space point of view suggested in this talk, the construction is "obviously" right in the local problem, while its validity in the global case is a delicate fact. The strategy described in the first three paragraphs after equation (1), and followed in this talk, will certainly work if A, B, and C are defined to be large enough spaces of functions. But if they are too small it could happen that ker g/im f would be not the whole fiber of E, but only a subspace. It is most clear that this does not happen if A, B, and C are defined without restriction as spaces of functions of four variables, as suggested in the comments just after equation (1). With the definitions actually given in this paper, A, B, and C being defined in terms of functions of three variables, it is slightly delicate but still relatively easy to see that the construction works. In the global problem treated in reference one, where the elements of A, B, and C are required not only to satisfy conditions (5), (6), and (7), but also to be global elements of the cohomology classes, these spaces become finite dimensional. In this case it is a quite delicate fact that the spaces A, B, and C are still large enough that the construction works.

Thus, this construction is also valid in a local form. As I mentioned at the outset of this talk, the reason that I think that this fact has some significance is related to the fact that the second order Minkowski space Yang-Mills equation, $D^{\mu}F_{\mu\nu} = 0$, is quite probably the equation we must come to grips with if we hope for applications to quantum field theory. These equations are hyperbolic; one can think of them in terms of a local problem or an initial value problem, but there is for them no global problem analogous to the global self-dual problem on S^4. The fact that the HBAHMD construction makes sense in a local version encourages one to hope that an analogue of this construction may exist for the second order Yang-Mills equations. The discovery of such an analogue would be very exciting.

A suitable starting point in a search for such a construction might be a construction recently found for the second order Yang-Mills equations by Yasskin, Isenberg, and Green [3] and by Witten [4]. This construction has been presented at this conference in talks by Yasskin and by Isenberg.

I would like to conclude with several technical remarks. In the local problem, A, B, and C are infinite dimensional vector spaces. The use of infinite dimensional spaces should not alarm or surprise us. It is analogous to the fact that in the inverse scattering appraoch to the sine-gordon equation, the sine-gordon field is reconstructed form linear maps among certain infinite dimensional vector spaces, the spaces of solutions of the auxiliary Schroedinger equation [5].

It should be noted that in the local problem, f is not injective, although g is still surjective. If one wants to find a sequence of spaces and maps exact except at one stage, the non-exactness corresponding to E, then it is necessary to introduce a fourth vector space; roughly speaking, the fourth space is the kernel of f. Also, it should be noted that certain simplifying features of this construction that appear in the global case are missing in the local case: A is not dual to C, B does not have a natural bilinear form, and g is not the adjoint of f (in the global problem these properties appear if E is symplectic or orthogonal).

The last point to be made here concerns the question of the choice of gauge groups or of the rank of vector bundles. Most discussion so far has concerned SU(2) gauge fields or rank two bundles. For applications to quantum field theory, however, the often made assumption that SU(2) is the simplest theory is probably an error.

Rather, it was shown by 't Hooft in a very important paper [6] that has not received all the attention it deserves that Yang-Mills theory with an SU(N) gauge group is, as a quantum field theory, simplest in the limit $N \to \infty$. I believe that the most realistic goal in this field is to try to understand the large N limit proposed by 't Hooft.

It is very striking that the HBAHMD construction also simplifies (for fixed value of the Chern class) as $N \to \infty$ [7]. This convergence of the region in which the quantum field theory is known to simplify and the region in which the mathematics of classical solutions simplifies is very intriguing, and perhaps of great significance. It may well turn out to be, in the long run, the most significant aspect of the construction, if one has physical applications in mind.

I would conjecture that a generalization of the HBAHMD construction to the second order equations, if it exists, will be something that is uselessly awkward for any finite N, and is tractable only at $N = \infty$.

References

1 G. Horrocks, unpublished;
W. Barth, unpublished;
M. Atiyah, N. Hitchin, Yu. I. Manin, and V. G. Drinfeld, Phys. Lett. 65A, 185(1978).
Yu. I. Manin and V. G. Drinfeld, Comm. Math. Phys., to be published.

2 R. Ward, Phys. Lett. 61A(1977) 81.

3 J. Isenberg, P. B. Yasskin, and P. S. Green, "Non Self-Dual Yang-Mills Fields," to be published in Phys. Lett. B.

4 E. Witten, "An interpretation of Classical Yang-Mills Theory," to be published in Phys. Lett. B.

5 Some additional comments on this analogy have been made by L. Fadde'ev (private communication).

6 G. 't Hooft, Nucl. Phys. B72(1974) 461.

7 M. Atiyah, private communication.

Acknowledgment

I would like to thank D. Kazhdan and D. Mumford for discussions.

E WITTEN
Lyman Physics Laboratories
Department of Physics
Harvard University
Cambridge, MA 02138

J Harnad, L Vinet and S Shnider

Solution to Yang-Mills equations on $\overline{M}^4$ invariant under subgroups of 0(4,2)

The purpose of this paper is to present, in outline, a systematic exposition of what is known about solutions to the Yang-Mills equations on compactified Minkowski space $\overline{M}^4$ which are invariant under a subgroup of the conformal group $O(4,2)$. If we identify $\overline{M}^4$ with $U(2)$ then the action of $O(4) \times O(2)$ on $\overline{M}^4$ is up to a covering map equivalent to the action of $SU(2)_L \times SU(2)_R \times U(1)$ where the subscript L or R indicates multiplication on the left or right respectively. We give a complete list of all solutions invariant under

a) $SU(2)_L \times SU(2)_R$

b) $SU(2)_L \times U(1)$

and all self dual solutions invariant under

c) $SU(2)_L$ (or $SU(2)_R$)

d) $(SU(2)_L \times SU(2)_R)_{diag.}$

First some notation:

Let

$$h = \begin{pmatrix} x_0+x_1 & x_2+ix_3 \\ x_2-ix_3 & x_0-x_1 \end{pmatrix} \in \mathrm{Herm}(2)$$

$$v = (I + ih)(I - ih)^{-1} \in U(2)$$

$$v = \begin{pmatrix} u_5+iu_0 & 0 \\ 0 & u_5+iu_0 \end{pmatrix} \begin{pmatrix} u_4+iu_1 & u_2+iu_3 \\ -u_2+iu_3 & u_4-iu_1 \end{pmatrix}$$

where $u_0^2+u_5^2 = u_1^2+u_2^2+u_3^2+u_4^2 = 1$ and the vector $(u_a)\in\mathbb{R}^6$ is determined up to sign. Then $h \mapsto v$ imbeds $\mathbb{R}^4$ in the compact manifold U(2) as an open dense set in such a way that the pseudo metric $\det(v^{-1}dv)$ is conformally related to the standard Minkowski pseudo metric $\det(dh)$. The correspondence between Cartesian coordinates (x_μ) and group coordinates (u_i) is given by

$$x_\mu = \frac{u_\mu}{u_4+u_5} \qquad \mu = 0, 1, 2, 3.$$

The coordinates (u_i) identify the two-fold cover $\tilde{M}^4$ of $\bar{M}^4$ with $SU(2) \times U(1)$. Further if $u_4 = \cos\phi$ and $u_5 = \cos\psi$, ϕ being part of a spherical coordinate system on $SU(2) \cong S^3$, ψ an angular coordinate on $U(1) \cong S^1$, and $t = x_0$, $r = \sqrt{x_1^2+x_2^2+x_3^2}$, then

$$t+r = \tan\left(\frac{\psi+\phi}{2}\right) \tag{1}$$

$$t-r = \tan\left(\frac{\psi-\phi}{2}\right).$$

A basis for the left-invariant forms is given by

$$\omega^0 = d\psi \tag{2}$$

$$\omega^i = 2(u_4du_i - u_idu_4 + u_jdu_k - u_kdu_j)$$ where (i,j,k) is a cyclic permutation of (1,2,3). Let $\{\sigma_j\}$ be the Pauli matrices and $\{X_j = \sigma_j/2i\}$ the corresponding basis for the Lie algebra su(2).

Any su(2) vector potential (local connection form) on $\tilde{M}^4$ can be written

$$\omega = A^i_j(u_a)X_i\omega^j+B^i(u_a)X_i\omega^0. \tag{3}$$

The associated field (curvature) is

$$\Omega = d\omega + \frac{1}{2}[\omega,\omega]. \tag{4}$$

The source-free Yang-Mills equations are

$$d{*}\Omega + [\omega, {*}\Omega] = 0 \tag{5}$$

where the * operation on two forms is defined by the pseudometric

$$(\omega^0)^2 - \frac{1}{4}\,((\omega^1)^2 + (\omega^2)^2 + (\omega^3)^2).$$

If the curvature form is self-dual or anti-self-dual

$$*\Omega = \pm i\Omega \tag{6}$$

the Yang-Mills equations are a consequence of the Bianchi identities. We are looking for solutions satisfying a condition of invariance up to gauge transformations of the form

$$(L_g^*\omega)_u = \mathrm{Ad}\rho(g,u)^{-1}(\omega_u) + \rho(g,u)^{-1}d_u\rho(g,u),$$

for $g \in G$, one of the groups listed above. Two theorems which are very helpful in reducing the problem are given below.

<u>Theorem 1</u>

Let $(g,u) \longmapsto f_g(u)$ be the left action of SU(2) on $\tilde{M}^4$. Given any principle SU(2) bundle E over $\tilde{M}^4$, $\pi : E \to \tilde{M}^4$, on which there exists a group action $(g,b) \longmapsto \tilde{f}_g(b)$ such that $\pi\tilde{f}_g(b) = f_g(\pi(b))$ there is a bundle isomorphism $\tau : E \to \tilde{M}^4 \times SU(2)$. Furthermore by the appropriate choice of τ (appropriate gauge) we can write

$$\tau\tilde{f}_g\tau^{-1}(u,h) = (f_g(u),h),\ (u,h) \in \tilde{M}^4 \times SU(2).$$

This means the local expression for the vector-potential is globally defined on $\tilde{M}^4$ and the gauge invariance function ρ can be chosen to be $\rho(g,u) \equiv e$.

Theorem 2

Let $(g,u) \to f_g(u)$ be the conjugation of SU(2) on $\tilde{M}^4$ (which is the induced action of the covering group of SO(3) acting on $\tilde{M}^4$); then in each equivalence class of SU(2) bundles $\pi: E \to \tilde{M}^4$ (indexed by the integers [5]) there exist exactly two inequivalent group actions satisfying $\pi(\tilde{f}_g(b)) = f_g(\pi(b))$. We can always find a trivialisation τ of E over $(U(1) - \{e\}) \times SU(2)$ such that either

$$\tau \tilde{f}_g \tau^{-1}(u,h) = (f_g(u),h)$$

or

$$\tau \tilde{f}_g{}^{-1}(u,h) = (f_g(u),gh).$$

In this case not all bundles with SO(3) action are trivial so the vector potential is not necessarily globally defined. The gauge function ρ can be reduced to either $\rho(g,u) \equiv e$ or $\rho(g,u) = g$. We can now give a table of results and references. (See following page.)

In order to study cases 1, 2 of the table we look at the full $SU(2)_L$ invariant equations

$$\frac{1}{2}\frac{d^2A_k}{d\psi^2} + \frac{1}{2}\left[\frac{dB}{d\psi}, A_k\right] + \left[B, \frac{dA_k}{d\psi}\right] + 2A_k - 3\sum_{p,q} \varepsilon_{kpq}[A_p,A_q] - 2\sum_m [A_m,[A_m,A_k]] + \frac{1}{2}[B,[B,A_k]] = 0 \tag{8a}$$

and

$$\sum_\ell \left([A_\ell, \frac{dA_\ell}{d\psi}] - [A_\ell,[A_\ell,B]]\right) = 0 \tag{8b}$$

where $A_k = A_k^i X_i$ and $B = B^i X_i$. In case 1 the $SU(2)_R$ invariance implies $A_k^i = f\delta_k^i$ and $B = 0$, therefore there is only one equation

Invariance conditions and known solutions

	Invariance group	Invariance equation	Form of ansatz in equation (3)	What is Known
1)	$SU(2)_L \times SU(2)_R$ (equivalent to $O(4)$ invariance)	a) $f^*_{(g,h)}\omega = \text{Ad}\, h^{-1}\omega$	$A^i_j(\psi) = f(\psi)\delta^i_j$, $B^i = 0$	All solutions [3][8][10][12][13]
		b) $f^*_{(g,h)}\omega = \omega$	$A^i_j = 0 \quad B^i(\psi)$	Solutions pure gauge
2)	$U(2)_L$	$f^*_g\omega = \omega$	A^i_j, B^i constant	All solutions [3][4][5][6]
3)	$SU(2)_L$	$f^*_g\omega = \omega$	$A^i_j(\psi) \quad B^i(\psi)$	All (anti) self-dual solutions [5] some non-self dual solutions [5][6]
4)	$(SU(2)_L \times SU(2)_R)_{diag}$ (equivalent to $O(3)$ invariance)	a) $f^*_{(g,g)}\omega = \text{Ad}g^{-1}\omega$	$A^i_j(\phi,\psi) \quad B^i(\phi,\psi)$	All (anti) self-dual solutions, some non-self-dual [5][7][9][11]§
		b) $f^*_{(g,g)}\omega = \omega$	$\omega = X_i(A^i(\phi,\psi)d\phi + B^i(\phi,\psi)d\psi)$	All solutions [5]

§ Spherically symmetric static solutions on M^4 are studied in [7][9][11][5].

$$\frac{1}{2} f'' + 2f(1-f)(1-2f) = 0.$$

Luscher [10] and Schechter [13] have given a complete analysis of this class. The equations can be solved explicitly using elliptic functions. The constant solutions f = 0, 1 are pure gauge but the solution $f = \frac{1}{2}$ is non-trivial and in another gauge had already been found by deAlfaro, Fubini and Furlan [3]. The self dual solution was analyzed by Rebbi [12]. See also [2]. In case 2, A^i_j and B^i are constant. We define $X = (X^i_j) = (A^i_k A^j_k)$ and equations (8) become

$$\begin{aligned} &X(1 + \mathrm{tr}X) \mp 3(\det X)^{\frac{1}{2}} I - X^2 + \tfrac{1}{4}(B\otimes XB - B^2X) = 0 \\ &XB = (\mathrm{tr}X)B \end{aligned} \qquad (8')$$

where $B^2 = \Sigma B^i B^i$. Solving these equations for X, B and then looking for those A such that $X = A^t A$ and which also satisfy the original equations, one finds

Theorem 3

Up to constant gauge transformation the $U(2)_L$ invariant solutions to (8) are

I. $A = \alpha\otimes\gamma \quad B = \beta \quad$ with $\alpha, \beta \in su(2)\otimes\mathbb{C}$, $\gamma \in su(2)^*$

where $\alpha.\beta = 0$ and $\beta^2 = 4$

either $\alpha^2 = 0$ or $\gamma^2 = 0$

II. $A = \begin{pmatrix} i & & \\ & i & \\ & & 3 \end{pmatrix} R \qquad B = 0$

where $R \in SO(3, \mathbb{C})$

III. $A = \lambda I$ $\qquad B = 0$

$\lambda = \frac{1}{2}, 1$

IV. $A = 0$ $\qquad B$ arbitrary

see Harnad et al. [5]. Howe and Tucker [6] and Tucker and Zakrewski [14] give partial results to this and the $SU(2)_L$ case (case 3).

In case 3 we look at the (anti) self-dual equations

$$\frac{dA_j}{d\psi} \mp 2iA_j + [B,A_j] \pm i \sum_{k,\ell} \varepsilon_{jk\ell}[A_k,A_\ell] = 0. \tag{9}$$

However if the vectors $\{A_i\}$ span a one dimensional subspace at each point, then by first gauge transforming to $B = 0$ the full equations (8) become linear and are solvable.

Next we transform the equation by

$$A_j = -\tilde{A}_j\, e^{\pm 2i\psi} \qquad B = \tilde{B}e^{\pm 2i\psi}$$

and introduce the complex variable $w = e^{\pm 2i\psi}$. Then

$$\frac{d\tilde{A}_j}{dw} + [B,\tilde{A}_j] = \frac{1}{2} \sum_{k,\ell} \varepsilon_{jk\ell} [\tilde{A}_k,\tilde{A}_\ell]. \tag{10}$$

If $\{\tilde{A}_i\}$ spans a two dimensional subspace we can solve (10) immediately; otherwise we introduce the matrix $(Y^i_j) = (\tilde{A}_i.\tilde{A}_j)$

$$\frac{dY}{dw} = (\det Y)^{1/2} I. \tag{11}$$

Let g be such that

$$Y = g.I + C$$

where C is a constant complex symmetric matrix with trace C = 0. Then g satisfies

$$\left(\frac{dg}{dw}\right)^2 = \det Y = g^3 + ag + b \tag{12}$$

where a, b are integration parameters for (11). We solve for g in terms of elliptic functions then determine Y by specifying the remaining parameters in C, which we assume to be in a canonical form (using complex conformal transformations). Finally, we look for $\tilde{A}$ such that $Y = \tilde{A}\tilde{A}^t$ and $\tilde{A}$ satisfies (10).

<u>Theorem 4</u>

The $SU(2)_L$ invariant solutions to equation (6) are as follows:

1) If $\{\tilde{A}_1, \tilde{A}_2, \tilde{A}_3\}$ span a 1-dimensional space at each point and B = 0

$$A_j = \gamma_j \alpha \qquad \alpha \in su(2)\otimes\mathbb{C}$$
$$\gamma_j \text{ scalars}$$

2) If $\{\tilde{A}_1, \tilde{A}_2, \tilde{A}_3\}$ span a 2-dimensional space at each point then up to a constant gauge transformation and a cyclic permutation

$$\tilde{A}_1 = \alpha$$
$$\tilde{A}_2 = \beta$$
$$\tilde{A}_3 = 2(\alpha\cos\theta + \beta\sin\theta)$$
$$\tilde{B} = 2(-\alpha\sin\theta + \beta\cos\theta)$$

where θ is a complex parameter and $\alpha, \beta \in su(2)\otimes\mathbb{C}$.

3) If the vectors $\tilde{A}_i$ are linearly independent, then three classes of solutions to eq. (5.6) exist, depending upon whether the matrix Y has two (and hence three) eigenvectors of non-zero length, only one, or none at all. Denoting by $\tilde{A}$ the matrix whose columns are $\tilde{A}_1$, $\tilde{A}_2$, and $\tilde{A}_3$, then up to a gauge transformation, B vanishes and

(i) If Y has three eigenvectors of non-zero length

$$\tilde{A} = \begin{bmatrix} p & & \\ & q & \\ & & r \end{bmatrix} R$$

$$\left.\begin{aligned} \text{where } p &= b\,\mathrm{ds}\,(b(w-w_0) \mid m) \\ q &= b\,\mathrm{ns}\,(b(w-w_0) \mid m) \\ r &= b\,\mathrm{cs}\,(b(w-w_0) \mid m) \end{aligned}\right\}$$

ds, ns and cs are Jacobi-Glaisher functions, b, m, $w_0 \in \mathbb{C}$ and $R \in SO(3,\mathbb{C})$ are arbitrary.

(ii) If Y has only one eigenvector of non-zero length,

$$\tilde{A} = \begin{bmatrix} p+q & iq & 0 \\ iq & p-q & 0 \\ 0 & 0 & r \end{bmatrix} R$$

where

$$\left.\begin{aligned} p &= \frac{a}{\mathrm{sh}[a(w - w_0)]} \\ q &= \frac{b\,\mathrm{sh}[a(w - w_0)]}{a} \\ r &= \frac{b\,\coth[a(w - w_0)]}{a} \end{aligned}\right\}$$

with $a, b, w_0 \in \mathbb{C}$, $ab \neq 0$ and $R \in SO(3,\mathbb{C})$

(iii) If Y has no eigenvectors of non-zero length,

$$\tilde{A} = \begin{bmatrix} p+q & iq & r \\ iq & p-q & ir \\ r & ir & p \end{bmatrix} R$$

where

$$\left.\begin{aligned} p &= \frac{1}{w - w_0} \\ q &= -\frac{1}{2}(w - w_0)[a^2(w - w_0)^2 - b] \\ r &= -a(w - w_0) \end{aligned}\right\}$$

with a, b, $w_0 \in \mathbb{C}$, $a \neq 0$ and $R \in SO(3,\mathbb{C})$.

In case 4 with invariance conditions a) we write
$\omega = A\,\underset{\sim}{u}\,\omega^0 + B\underset{\sim}{\omega} + C\,\underset{\sim}{u}X\underset{\sim}{\omega} + D\,\underset{\sim}{u}\,(\underset{\sim}{u}.\underset{\sim}{\omega})$ where $\underset{\sim}{u} = (u_1, u_2, u_3)$ and $\underset{\sim}{\omega} = (\omega^1, \omega^2, \omega^3)$ and A, B, C, D are scalar functions of ϕ, ψ.

When $(B - \frac{1}{2})^2 + E^2/\underset{\sim}{u}^2 \neq 0$ where $E = u_4 + 2\underset{\sim}{u}^2\,C$, we can find a gauge in which $B = \frac{1}{2}$. When $E \neq 0$ we can write the equations as

$$\begin{aligned} &A = \pm i\,\frac{E_\phi}{E\sin\phi} \qquad D = \pm\, i\,\frac{E_\psi}{2\underset{\sim}{u}^2 E} \\ &\left(\frac{E_\psi}{E}\right)_\psi - \left(\frac{E_\phi}{E}\right)_\phi = \frac{1 - E^2}{\sin^2\phi}\,. \end{aligned} \tag{13}$$

Let $E = e^\rho\sin\phi$ and the last equation becomes

$$\rho_{\phi\phi} - \rho_{\psi\psi} = e^{2\rho}\,.$$

Let $\xi = \frac{\psi+\phi}{2}$ $\quad \eta = \frac{\psi-\phi}{2}$ and we have

$$\rho_{\xi\eta} = -e^{2\rho}$$

which has a general solution

$$\rho(\xi,\eta) = \ln \left(\frac{(F'(\xi)G'(\eta))^{1/2}}{F(\xi)-G(\eta)}\right)$$

$$E^2 = \frac{F'(\xi)G'(\eta)}{(F(\xi)-G(\eta))^2} \quad \sin^2(\xi-\eta). \tag{14}$$

In order to have regularity at the spatial origin $\phi = 0$ ($\xi=\eta$) we must have F=G and both of period π. Write

$$F(\xi) = f(\tan\xi) \qquad G(\eta) = g(\tan\eta)$$

and transform to cartesian coordinates; the generating function E becomes

$$E(r,t) = \frac{f'(r+t)g'(t-r)}{(f(r+t)-g(t-r))^2} \cdot r^2$$

The resulting solutions are regular on M^4 provided f=g and is strictly monotonic, however the solutions will not be regular on the compactified space $\bar{M}^4$.

The SO(3) invariant self-dual solutions in Euclidean space were first studied by Witten [15] and his analysis carried over to Minkowski space is equivalent to ours under a suitable change of gauge and basis. For some particular spherically symmetric static solutions on $\bar{M}^4$ see Hsu and Mac [7], Ju [9] and Protogenov [11].

References

1 W. Bernreuther, Phys. Rev., D 16, 3609, (1977).

2 J. Cervero, L. Jacobs, C. Hohl, Phys. Lett., 69B, 351, (1977).

3 V. deAlfaro, S. Fubini, G. Furlan, Phys. Lett., 65 B, 163, (1976). and CERN preprint TH2397 Oct., 1977.

4 J. Harnad, L. Vinet, Phys. Lett., 76 B, 589, (1978).

5 J. Harnad, L. Vinet, S. Shnider, CRM preprint 792 (submitted to J. Math. Phys.), part-two in preparation.

6 P. Howe, R. Tucker, CERN preprint TH2421 (to appear in Nucl. Phys.)

7 J. P. Hsu and E. Mac, J. Math. Phys., 18, 100(1977).

8 R. Jackiw, C. Nohl, C. Rebbi, "Classical and semi-classical solutions of the Yang-Mills Theory" in Boal and Kamal, Particles and Fields, Plenum Press 1978.

9 I. Ju., Phys. Rev., D 17, 1637 (1978).

10 M. Luscher, Phys. Lett., 70 B, 321 (1977).

11 A. P. Protogenov, Phys. Lett., 67 B, 62 (1977).

12 C. Rebbi, Phys. Rev., D 17, 483 (1978).

13 B. Schechter, Phys. Rev., D 16, 3015 (1977).

14 R. Tucker, W. Zakrewski, CERN preprint TH 2462.

15 E. Witten, Phys. Rev. Lett., 38, 121 (1977).

Acknowledgment

This research was supported by NRC grants.

J HARNAD and L VINET
Centre des Recherches Mathematiques
Universite de Montreal
Montreal, Canada

and

S SHNIDER
Department of Mathematics
McGill University
Montreal, Canada

P Green

Integrality of the Coulomb charge in the line space formalism

We will consider the Coulomb field as the curvature of a connection on a trivial complex line bundle whose base space is the complement of the world line of a point charge, taken to be the t-axis in Minkowski space. Having chosen an appropriate background trivialization, we may identify the connection with the one-form $\tilde{q}dt/r$. Here $r^2 = x^2 + y^2 + z^2$, and $\tilde{q}$ is an appropriate multiple of the electric charge.

As long as we confine our attention to the field on real Minkowski space, $\tilde{q}$ may take on any real value. What we will show in the present note is that if we complexify the field and require that the resulting complex field be obtainable, as in [1], from a bundle over the space of null lines in its domain, it follows that $\tilde{q}$ must be an integer.

In order to interpret $\tilde{q}dt/r$ as a complex field, we take as its domain, $V = \{x,y,z,t,r) | r^2 = x^2 + y^2 + z^2 \neq 0\}$. Since r has exactly two possible values for each choice of (x,y,z,t), V is a double covering of $C^4 - Q$ where $Q = \{(x,y,z,t) | x^2 + y^2 + z^2 = 0\}$. This is necessary in order to make sense of r.

The hypothesis that the field is obtainable from a bundle over the space of null lines in V is equivalent to the existence of a globally covariantly constant section of the bundle over every null line. Let us consider the line $x = t = \lambda$, $y = z = 0$. This line intersects Q only in the origin. It is covered in V by two copies on which $r = \lambda$ and $r = -\lambda$ respectively. Restricting our attention to the copy on which $r = \lambda$, we may write $\tilde{q}dt/r = \tilde{q}d\lambda/\lambda$. The parallel translation law may be described as follows:

Let ϕ be a section which is constant in the background trivialization. Then the parallel translation of $\phi(\gamma(0))$ along the curve γ from $\gamma(0)$ to $\gamma(1)$ may be written $\exp(\int_\gamma \tilde{q} d\lambda/\lambda)\phi(\gamma(1))$. If there exists a globally covariantly constant section over the line, then parallel translation around the line must be independent of path. In particular it must be the case that $\exp(\int_\gamma \tilde{q} d\lambda/\lambda) = 1$ whenever γ is a closed path. It follows that, for every closed path γ, $\int_\gamma \tilde{q} d\lambda/\lambda$ is an integral multiple of $2\pi i$. Since $\int_\gamma \tilde{q} d\lambda/\lambda$ is evidently $2\pi i\tilde{q}$ when γ is any closed curve which encircles the origin once, we deduce that $\tilde{q}$ is an integer.

Although the general null line is more complicated (see remark 2 below), it can be verified that the integrality of $\tilde{q}$ is a sufficient condition for global integrability on each null line.

Remarks

1. Until the precise relation between $\tilde{q}$ and the electric charge is clarified, the physical implications of the integrality of $\tilde{q}$ are not clear.
2. The space of null lines of V is a non-Hausdorff manifold owing to the fact that a generic line in C^4 meets Q in two points and is non-trivially covered by its preimage in V, while there is an exceptional set of lines which meet Q in a single point or (even more exceptionally) not at all. Each of these exceptional lines is trivially covered and its preimage consists of two null lines of V instead of one. The geometry of this line space is very interesting and I hope to return to it in another paper.
3. The necessity of passing to a twofold covering of $C^4 - Q$ is somewhat obnoxious, particularly since it raises the possibility of passing to a still higher covering, which would rob the result of all its force. (This may, of course, be desirable in view of Remark 1 above.) Two points can

be raised in this connection. The first is that higher coverings would be of a different geometrical nature from the one that defines V; only the extremely exceptional lines which miss Q altogether would be trivially covered. The second point is that the introduction of V is less ad hoc if one reflects on the fact that line bundles on V correspond in a natural way to a class of O(2,C) bundles on $C^4 - Q$. This suggests that one should take the background of the Coulomb connection to be a non-trivial O(2,C) bundle (which becomes trivial when lifted to V) rather than a trivial line bundle.

References

1 Yasskin, Isenberg, Green, Non-Self-Dual Gauge Fields. Phys. Letters B, to appear.

PAUL GREEN
Department of Mathematics
University of Maryland
College Park, Maryland 20742

N H Christ

Unified weak and electromagnetic interactions, baryon number nonconservation and the Atiyah-Singer theorem

There are two quite distinct classes of physical phenomena which are currently believed to be described by a Yang-Mills theory: 1. The strong interactions (nuclear forces) where the quanta of the gauge field and the isolated "charges" to which it couples are imagined to be unobservable (confinement). 2. Weak and electromagnetic interactions. Here the photon is an honest-to-goodness gauge particle and the other gauge quanta whose exchanges mediate the weak interactions (e.g., β-decay or neutrino scattering) are very massive and as yet unobserved. However, it is widely expected that examples of these particles will be discovered within the next decade.

In this talk I would like first to describe the simplest model of such a Yang-Mills theory of weak and electromagnetic interactions, the Weinberg-Salam model [1]. Second, following 't Hooft [2], I will discuss the relationship between the Pontryagin index, the Adler anomaly and the nonconservation of baryon number. Finally I will show how the Atiyah-Singer theorem implements this nonconservation of baryon number in a semiclassical, Euclidean-space calculation.

The Weinberg-Salam model is built upon the gauge group SU(2) × U(1). In addition to the four gauge fields $\{A^i_\mu\}_{1\le i\le 3}$ and B_μ corresponding to the generators of SU(2) × U(1) this model contains a doubtlet of scalar fields $\{\phi^\alpha(x)\}_{\alpha=1,2}$ belonging to the fundamental representation of SU(2), a number $N_L \ge 12$ of similar doublets of left-handed Fermi fields $\{\psi^\alpha_{Li}(x)\}_{1\le i\le N_L}$ and N_R right-handed Fermi fields $\{\psi_{Ri}(x)\}_{1\le i\le N_R}$, unaffected by SU(2) transformations. In addition to the electromagnetic field $A_\mu(x)$ which is a linear combination of the gauge fields A_μ^3 and B_μ only the Fermi fields ψ_i^α correspond to known particles--the "known" quarks and leptons.

The gauge fields obey the following equations [3]

$$[\partial_\mu+g\vec{A}_\mu\times][(\partial^\mu\vec{A}^\nu-\partial^\nu\vec{A}^\mu)+\vec{A}^\mu\times\vec{A}^\nu]$$

$$= -g\sum_{i=1}^{N_L}\overline{\psi}^\alpha_{L,i}\gamma^\nu(\vec{\tau}^\alpha_\beta/2)\psi^\beta_{L,i} + g\phi^{\alpha\dagger}[\tfrac{1}{2}(-i\overleftrightarrow{\partial}^\nu+g'B^\nu)\vec{\tau}^\alpha_\beta - \tfrac{g}{2}\vec{A}^\nu\delta^\alpha_\beta]\phi^\beta \tag{1a}$$

$$\partial_\mu[\partial^\mu B^\nu-\partial^\nu B^\mu] = -g'\sum_{i=1}^{N_L} y_i^L\overline{\psi}^\alpha_{L,i}\gamma^\nu\psi^\alpha_{L,i} - g'\sum_{i=1}^{N_R} y_i^R\overline{\psi}^\alpha_{R,i}\gamma^\nu\psi^\alpha_{R,i}$$

$$- g'\phi^{\alpha\dagger}[\tfrac{1}{2}(-i\partial^\nu+g'B^\nu)\delta^\alpha_\beta-g\vec{A}^\nu(\tau^\alpha_\beta/2)]\phi^\beta \tag{1b}$$

where y_i^L and y_i^R are simple fractions chosen so that the quarks and leptons have the proper electric charges, $\{(\tau^i)^\alpha_\beta\}_{1\le i\le 3}$ are the three 2×2 Pauli matrices and we have used a vector notation to represent the three SU(2) indices of A_μ^i. If all terms non-linear in the fields are dropped from Eq.(1) then only derivatives of the fields A_μ^i and B_μ^i appear. This would imply that the corresponding four quanta are massless particles--a physically untenable situation. Consequently, the Weinberg-Salam model postulates a self-interaction of the field ϕ^α so arranged that in the configuration of minimum energy the field $\phi^\alpha(x)$ is a non-vanishing constant. Thus, if we choose $\phi^1(x) = 0$, $\phi^2(x) = v/\sqrt{2}$, substitute into Eq. (1) and then drop all non-linear terms, the resulting equations are

$$\partial_\mu[\partial^\mu\vec{A}^\nu-\partial^\nu\vec{A}^\mu] = -\tfrac{1}{4}g[g\vec{A}^\nu+g'\hat{e}_3B^\nu]|v|^2$$

$$\partial_\mu[\partial^\mu B^\nu-\partial^\nu B^\mu] = -\tfrac{1}{4}g'[g(A^3)^\nu+g'B^\nu]|v|^2 \tag{2}$$

where $\hat{e}_3$ is a unit vector in the three direction. The non-derivative terms on the right hand side of Eq. (2) give the corresponding quantum mechanical particles a mass: the quanta of A^1 and A^2 have mass $\frac{1}{2}g|v|$ the quanta of $gA^3+g'B$ have mass $\frac{1}{2}\sqrt{g^2+g'^2}|v|$ while those of $-g'A^3+gB$ have mass zero--a physically acceptable spectrum if $g|v| \gtrsim 50$ GeV.

As observed by 't Hooft, the conservation law for the baryon number current in this model contains an Adler anomaly [4]:

$$\partial_\mu j^\mu_B = -\frac{N_B}{3}\frac{g^2}{32\pi^2} F^i_{\mu\nu}(\tilde{F}^i)^{\mu\nu} \tag{3}$$

where N_B is the number of doublets of left-handed quark fields, $F^i_{\mu\nu}$ is the field strength tensor

$$F^i_{\mu\nu} = \partial_\mu A^i_\nu - \partial_\nu A^i_\mu + g(\vec{A}_\mu \times \vec{A}_\nu)^i \tag{4}$$

and $\tilde{F}^i_{\mu\nu}$ its dual. The baryon number j^μ_B is given by

$$j^\mu_B = \frac{1}{3}\sum_i \overline{\psi}_i\gamma^\mu\psi_i, \tag{5}$$

the sum extending over all right- and left-handed quark fields. The quarks carry baryon number ±1/3 and are the only particles with non-zero baryon number. The Adler anomaly (i.e., the right hand side of Eq. (3)) is a quantum mechanical effect and reflects the slow convergence of sums over large-momentum quantum states. The anomaly is not present in the classical field theory where baryon number is exactly conserved.

Since the time component $j^0_B(x)$ is the baryon number density, the increase in total baryon number Q_B between the times t_i and t_f is

$$\begin{aligned} Q_B(t_f) - Q_B(t_i) &= \int_{t_i}^{t_f} dt\, \partial_0 \int j^0_B(\vec{x},t)d^3x \\ &= \int_{t_i}^{t_f} \int \partial_\mu j^\mu_B d^3x\, dt \end{aligned} \tag{6}$$

where a surface integral at spacial infinity has been dropped. Thus in the limit $t_f \to +\infty$, $t_i \to -\infty$ the change in baryon number is $-N_B\nu/3$ where ν is the usual Pontryagin index for the Yang-Mills field. Consequently, if gauge field configurations with non-vanishing Pontryagin index play a role in the quantum mechanics, one expects a violation of baryon number in the Weinberg-Salam model and a resulant instability of nuclear matter!

Of course such an observation is interesting only to the extent that one can actually calculate the size of the effect. The following computational scheme has been developed by 't Hooft. One begins with a functional integral expression for a Green's function containing 2n Fermi fields:

$$\begin{aligned} G(x_1,\ldots,x_{2n}) \propto & \int d[A_\mu^i]\, d\,[B_\nu]\, d\,[\phi^\alpha]\, d\,[\bar{\psi}]\, d\,[\psi] \\ & \times \exp\{-i\int [L_{YM}+L_H+L_F] d^4x\} \\ & \times [\psi_{i_1}(x_1)\ldots\psi_{i_n}(x_n)\bar{\psi}_{i_{n+1}}(x_{n+1})\ldots\bar{\psi}_{i_{2n}}(x_{2n})] \end{aligned} \tag{7}$$

Here $\int L_{YM} d^3x$, $\int L_H d^3x$, and $\int L_F d^3x$ are the Lagrangians for the Yang-Mills fields, the scalar field and the Fermi fields respectively.

Perhaps the least familiar part of the "Feynman path integral" (7) is the integration over the Fermi fields ψ_i. If we treat the gauge and scalar fields as fixed, then we can expand each Fermi field $\psi(x)$ in terms of the eigenstates $\chi_\ell(x)$ of the Dirac operator

$$\begin{aligned} D = & i\gamma^\mu[\partial_\mu - \frac{g}{4}\vec{\tau}\cdot\vec{A}_\mu(1-\gamma^5) - i\frac{g'}{2}y_i^L B_\mu(1-\gamma_5) - i\frac{g'}{2}y_i^R B_\mu(1+\gamma_5)] \\ & - h_{ij}\phi - h'_{ij}\phi^\dagger \end{aligned} \tag{8}$$

which appears in L_F:

$$\psi(x) = \sum_\ell q_\ell \chi_\ell(x). \tag{9}$$

The Dirac matrices $(1 \pm \gamma_5)/2$ are projection operators onto right/left-handed states. In terms of the q_ℓ's the Fermi action becomes

$$\int L_F d^4x = \int \bar{\psi} D \psi d^4x = \sum_\ell \lambda_\ell q_\ell^\dagger q_\ell \tag{10}$$

where λ_ℓ is the eigenvalue of the operator (8) corresponding to χ_ℓ. The functional integral over the Fermi field $\psi(x)$ is then replaced by a multiple integral over the coefficients q_ℓ and $q_\ell^\dagger$. These are assumed to be anti-commuting algebraic quantities and all integrals computed according to the rules [5]

$$\int dq_\ell = \int dq_\ell^\dagger = 0 \tag{11a}$$

$$\int dq_\ell \, q_\ell = \int dq_\ell^\dagger \, q_\ell^\dagger = 1. \tag{11b}$$

Note higher powers of q_ℓ automatically vanish since $q_\ell \, q_\ell = -q_\ell \, q_\ell$ because of their anticommutation.

In order to compute low energy, baryon number violating contributions to the Green's function (7) one begins in Euclidean space and continues to physical Minkowski space only after the amplitude has been completely evaluated. The Euclidean integral over A^i, B and ϕ^α is done by a saddle point method. If the scalar field were absent the extrema of the action, $\int L_{YM} d^4x$, with Pontryagin index $\nu = 1$ would be configurations with $B_\mu = 0$ and A_μ^i a one-instanton solution with position z and scale parameter ρ. The inclusion of the scalar field ϕ^α with the self-interaction required by the Weinberg-Salam model spoils these extrema but in a very simple way. If the gauge coupling g is very small compared to the self interactions of ϕ^α, then the one instanton solution is extremal with respect to all variations except that of ρ. The action is an increasing function of ρ and only for very small instantons, $\rho << 1/|v|$, does it become the usual $8\pi^2/g^2$. There are no exact classical solutions to the coupled Yang-Mills and scalar field equations.

This ρ dependence of the action is not a serious obstacle--it is possible to perform one-dimensional integrals which are not Gaussian. In fact it is an enormous assistance because configurations of large, overlapping instantons (which obscure the application of these methods in the case of the strong interactions) are now highly suppressed.

The baryon number conservation equation (3) implies that for a Yang-Mills one-instanton configuration like that above, the resulting Green's function should violate baryon number by $N_B/3$ units. Similarly a configuration with Pontryagin index ν must contribute to a Green's function violating baryon number by $N_B\nu/3$ units. This relationship between the Pontryagin index ν and the above Green's function calculation is directly implied by the Atiyah-Singer index theorem. In order to understand this we must be somewhat more specific about the nature of the Fermi field $\psi(x)$ appearing in Eq. (10) when we continue to Euclidean space. In Minkowski space that part of the action which connects the SU(2) gauge field and Fermi fields in the two dimensional representation of SU(2),

$$\sum_{i=1}^{N_L} \int d^4x\, \bar{\psi}_{L,i}\, i\gamma_\mu(\partial_\mu - i\,\frac{g}{2}\,\vec{\tau}\cdot\vec{A}_\mu)\psi_{L,i}, \qquad (12)$$

can consistently be written in terms of a left-handed doublet $\psi^\alpha_{L,i}$ and its hermitian conjugate. However, in Euclidean space, the Dirac matrix γ_μ connects left-handed fields with the hermitian conjugates of right-handed fields so that the ψ in Eq. (12) must be a full four-component Fermi field containing both left-and right-handed fields. However, since the Weinberg-Salam model always contains an even number of left-handed Fermi fields with couplings to the SU(2) gauge field this is always possible. One must simply use for one half these Fermions the charge conjugate fields $\psi^c_{R,i} = \gamma_2\psi^\dagger_{L,i}$ in which the roles of particles and anti-particles are reversed and which are right-handed [6].

We can now ask for the consequences of the Atiyah-Singer theorem. In the approximation that g is small compared to the scalar self-interaction, the B_μ and ϕ terms in Eq. (10) can be ignored. Thus the action becomes a sum over $N_L/2$ identical terms

$$\sum_i^{\frac{1}{2}N_L} \int d^4x\, \overline{\psi}_i\, i\gamma_\mu(\partial_\mu - i\,\frac{g}{2}\,\vec{A}_\mu\cdot\vec{\tau})\psi_i \tag{13}$$

where each ψ_i is now a four-component Dirac field made up of left- and right-handed parts as described above. The Atiyah-Singer theorem implies that the operator $i\gamma_\mu(\partial_\mu - i\,\frac{g}{2}\,\vec{A}_\mu\cdot\vec{\tau})$ has ν vanishing eigenvalues. The corresponding eigenstates are left-handed if ν is positive and right-handed if ν is negative. Let $\{q_{i,\ell}\}$ $1 \leq \ell \leq \nu$ be the coefficients of these ν eignestates for the ith field ψ_i, $1 \leq i \leq \frac{1}{2}N_L$. Since the Euclidean Dirac matrix γ_μ connects left- and right-handed states, positive ν implies that the $q_{i,\ell}$ are coefficients of left-handed eigenstates while $q^\dagger_{i,\ell}$ appears in the right-handed part of $\overline{\psi}_i$. Because of the integration rules (11a) these variables $q_{i,\ell}$ and $q^\dagger_{i,\ell}$, being absent from $\int \mathcal{L} d^4x$, must appear in the explicit fields $\psi_{i_1}(x_1)\ldots\overline{\psi}_{i_{2n}}(x_{2n})$ in the integrand of Eq. (7) if the contribution to $G(x_1,\ldots,x_n)$ is not to vanish. Thus integration over the $q_{i,\ell}$ and $q^\dagger_{i,\ell}$ requires the presence of $\nu N_B/2$ left-handed fields $\psi_{L,i}$ with baryon number 1/3 and $\nu N_B/2$ right-handed fields $\psi^{c\dagger}_{R,i}$ also with baryon number 1/3. Since each such operator represents the annihilation of a particle carrying baryon number 1/3, the corresponding Green's function describes a process in which a total baryon number of $N_B\nu/3$ disappears--exactly as predicted by Eq.'s (3) and (6). (The integration over the remaining q's and $q^\dagger$'s is sufficiently symmetrical that no additional baryon number violation results.)

Because of the presence of the precise number of eigenstates with vanishing eigenvalue implied by the Atiyah-Singer theorem, this Euclidean space calculation agrees nicely with the baryon number violation predicted by Eq. (6). However, our Euclidean space calculation also determines the size of the effect. For $N_B = 6$ the minimal baryon number change is 2 when $\nu = 1$. The factor $\exp[-\int L_{YM} d^4x]$ for a one Euclidean instanton configuration in the Weinberg-Salam model becomes $\exp[-\frac{2\pi}{\alpha}\sin^2\theta_W]$ where $\sin^2\theta_W = g'^2/(g^2 + g'^2) \simeq 1/4$ and the fine structure constant $\alpha \simeq 1/137$, giving an overall factor of e^{-215}. This factor is sufficiently small that very few deuterons have decayed since the universe was created! Of course we might speculate that the complete theory of "weak and electromagnetic" phenomona might require a larger gauge group in which instantons of smaller action can occur, allowing predictions of the type just discussed but of greater physical interest.

References

1 S. Weinberg, Phys. Rev. Lett. 19, 1264(1967);
A. Salam, in Elementary Particle Physics, N. Svartholm, ed. (Almquist and Wiksells, Stockholm, 1968) p. 367.

2 G. 't Hooft, Phys. Rev. Lett. 37, 8(1976); Phys. Rev. D14, 3432 (1976).

3 We use the notation of Bjorken and Drell, Relativistic Quantum Fields (mcGraw-Hill, New York, 1965).

4 S. L. Adler, Phys. Rev. 177, 2426(1969).

5 F. A. Berezin, The Method of Second Quantization (Academic Press, New York, 1966).

6 The author thanks Savas Dimopolous for pointing this out to him.

7 A. Schwartz, Phys. Lett. 67B, 172(1977);

M. F. Atiyah, N. J. Hitchen and I. M. Singer, Proc. Natl. Acad. Sci. USA 74, 2662 (1977).

L. S. Brown, R. D. Carlitz and C. Lee, Phys. Rev. D 16, 417(1977).

Acknowledgment

This research was supported in part by the United States Department of Energy.

N H CHRIST
Department of Physics
Columbia University
New York, New York 10027